Letts

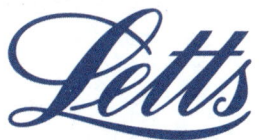

A-level In a Week

Chemistry

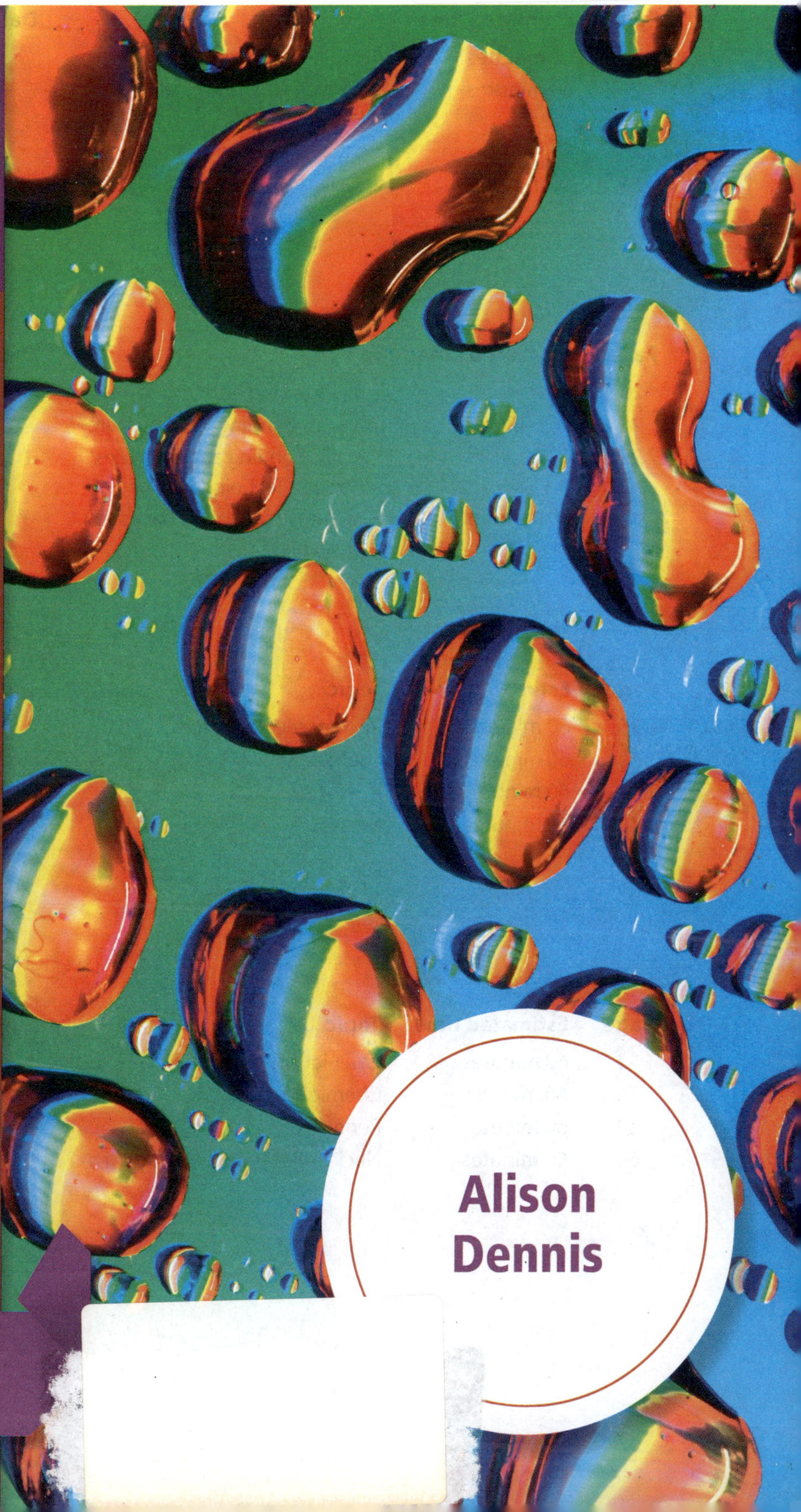

Year 1 / AS

Alison Dennis

CONTENTS

DAY 1

Page	Estimated time	Topic	Date	Time taken	Completed
4	60 minutes	Atomic Structure			☐
8	60 minutes	Representing Chemical Reactions			☐
12	60 minutes	Electron Configuration			☐
16	60 minutes	Calculations 1			☐

DAY 2

Page	Estimated time	Topic	Date	Time taken	Completed
20	60 minutes	Ionic Bonding and Structure			☐
24	60 minutes	Covalent Bonding and Structure			☐
28	60 minutes	Metallic Bonding and Structure/Titration Techniques			☐
32	60 minutes	Structures of Carbon/Shapes of Molecules			☐

DAY 3

Page	Estimated time	Topic	Date	Time taken	Completed
36	60 minutes	Intermolecular Forces			☐
40	60 minutes	Energetics			☐
44	60 minutes	Hess's Law			☐
48	60 minutes	Kinetics			☐

DAY 4

Page	Estimated time	Topic	Date	Time taken	Completed
52	60 minutes	Calculations 2			☐
56	60 minutes	Chemical Equilibria			☐
60	60 minutes	Redox			☐
64	60 minutes	Nuclear Reactions & Radiation			☐

DAY 5

Page	Estimated time	Topic	Date	Time taken	Completed
68	60 minutes	Periodicity/Group 2 Elements			☐
72	60 minutes	Group 7: The Halogens			☐
76	60 minutes	Uses of Group 2 and 7 Elements and Compounds			☐
80	60 minutes	Chemistry and the Environment			☐

DAY 6

Page	Estimated time	Topic	Date	Time taken	Completed
84	60 minutes	Organic Chemistry			☐
88	60 minutes	Alkanes			☐
92	60 minutes	Haloalkanes			☐
96	60 minutes	Alkenes			☐

DAY 7

Page	Estimated time	Topic	Date	Time taken	Completed
100	60 minutes	Alcohols			☐
104	60 minutes	Experimental Techniques			☐
108	60 minutes	Mass Spectrometry			☐
112	60 minutes	Infrared Spectroscopy			☐

116	Answers
130	Periodic Table
132	Index

DAY 1 — 60 Minutes

Atomic Structure

A Chemist's Model of the Atom

Atoms consist of a very small nucleus containing positive protons and neutral neutrons surrounded by negative electrons in shells or energy levels, e.g. lithium: 3 protons and 4 neutrons in the nucleus, 3 electrons surrounding the nucleus.

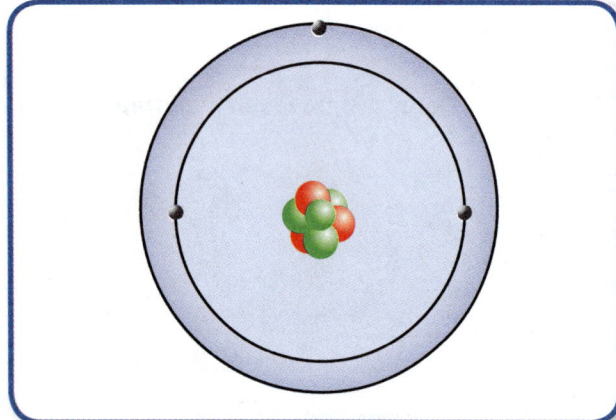

Most of the volume of the atom is made up of the electron shells (the diagram shows an enlarged nucleus to illustrate the particles). Each electron shell is divided into sub-shells of differing energy. Each sub-shell consists of varying numbers of orbitals, which can contain up to two electrons.

Properties of Sub-atomic Particles

Particle	Relative mass	Relative charge
proton	1	+1
neutron	1	0
electron	$\frac{1}{1840}$	−1

Each element has a unique atomic number, Z, which is equal to the number of protons in each atom of that element.

The number of neutrons can vary for atoms of the same element. Atoms with the same number of protons but different numbers of neutrons are **isotopes**.

The number of electrons in a neutral *atom* is the same as the number of protons.

The chemical properties of an element are mainly determined by the electrons in the outer shell. Elements with the same number of outer shell electrons are chemically similar.

Atoms may lose or gain electrons to form ions but the atom remains the same element because it has the same atomic (proton) number.

The number of electrons in an ion is one less than the atomic number for every positive charge on the ion and one more than the atomic number for every negative charge on the ion.

The mass number of an atom is the number of protons plus the number of neutrons.

Examples of sub-atomic particles:

Particle	Protons	Electrons	Neutrons
Pt	78	78	117
O^{2-}	8	10	8
Na^+	11	10	12

Representing Atoms

Letters represent the name of the element.

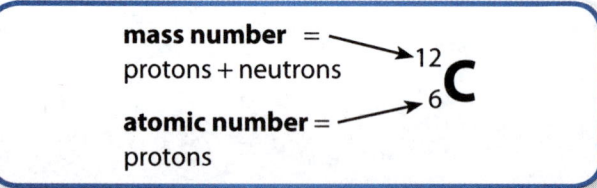

The number of neutrons in the atom can be calculated by mass number − atomic number.

Isotopes of the same element contain the same number of protons but a different number of neutrons.

E.g. isotopes of carbon $^{13}_{6}C$ $^{14}_{6}C$

Isotopes of an element have the same chemical properties but different physical properties, e.g. density and melting point.

Isotopes are named using their mass number, e.g. carbon-12, carbon-13, carbon-14.

Mass of Atoms

The standard mass unit is an atom of carbon-12, which has a value of 12. All other atomic, formula and molecular masses are quoted relative to this.

Relative Isotopic Mass is the relative isotopic mass of an atom compared with $\frac{1}{12}$ mass of an atom of carbon-12.

It is the same as the mass number of an atom of that isotope.

Relative Atomic Mass, A_r, is the weighted mean mass of an element compared to $\frac{1}{12}$ mass an atom of carbon-12.

This takes into account the relative abundance of each isotope of the atom.

The number given on the periodic table is the relative atomic mass.

(For how to calculate relative atomic mass see page 108 Mass Spectrometry.)

Development of Theories about Atoms

The model of the atom has changed over time as technology has improved and more evidence has been collected.

Atoms as the Tiny Particles from which the Universe is Made

Everything is made of many different atoms, which themselves are physically indivisible.

Evidence: Observations of the way that substances can be broken down and built up again.

Key 'scientist': Democritus ≈ 400 BC

Billiard Ball Model   WRONG

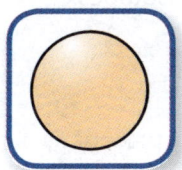

Atoms are the indivisible particles of elements. The atoms of each element have their own characteristic mass. In chemical reactions the atoms are rearranged.

Evidence: Experiments that identified elements as substances that cannot be broken down to anything simpler. Calculations of the atomic masses of different elements from careful experiments measuring the ratios in which they react.

Key scientist: John Dalton 1808

Plum Pudding Model   CLOSE

Atoms are not indivisible but made from positive and negative matter. Negative electrons are embedded in a positive matrix.

Evidence: **Negative** particles were emitted from the **neutral** atoms of a cathode (known as cathode rays) that could only have come from the atoms of the cathode. The particles had negative charge and very little mass. They were the same mass and charge no matter what the element they were emitted from. cool!

Key scientist: John Joseph Thompson 1897

Planetary Model   CLOSER

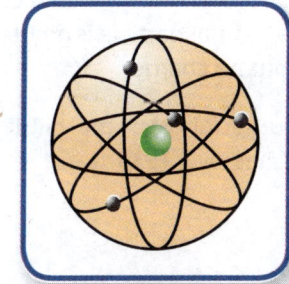

Atoms have a very small positive nucleus around which negative electrons orbit.

Evidence: α (alpha) particles fired at gold film mostly passed through without being deflected but were occasionally deflected by up to 180°. Mathematics demonstrated that this deflection can only happen if all the positive charge of an atom is concentrated into a very small area and not spread out as a matrix.

Key scientist: Ernest Rutherford 1911

Rutherford later showed that the charge on the nucleus is always a multiple of the charge on a hydrogen nucleus. He proposed that each element has a different number of positive particles (protons) in the nucleus of its atoms.

5

DAY 1

Quantised Shell Model

YES!

Electrons exist in shells of fixed energy around the nucleus.

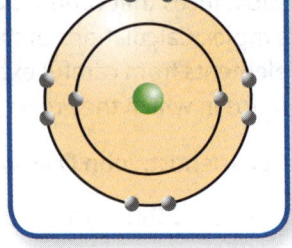

Evidence: Emission spectra. Atoms absorb and emit electromagnetic radiation of specific frequency. This is the result of the movement of electrons between different shells of fixed energy value.

The successive ionisation energies of an element show very large increases when an electron is removed from an inner shell.

Key scientist: Niels Bohr 1913

Structure in the Nucleus

The nucleus is composed of positively charged protons and neutral neutrons, which have equal mass.

Evidence: The atomic number of an atom is always proportional to the number of positive charges but not to its mass. When α particles are fired at beryllium, a neutral particle is emitted with the mass of a proton but no charge.

Key scientists: Ernest Rutherford and James Chadwick 1935

Quantum Model

Orbitals are regions where an electron is most likely to exist. Electrons can be considered to be waves. Electrons form new bonding orbitals when atoms react together.

Evidence: Mathematical calculations.

Key scientist: Erwin Schrödinger 1922

Don't forget

…a model is accepted until a better model is provided or new evidence is discovered that contradicts the current model.

QUICK TEST

1. Define relative isotopic mass of an atom.
2. Calculate how many neutrons an atom of carbon-14 contains.
3. How many electrons are in a Ca^{2+} ion?
4. What is the relative mass of a proton?
5. What evidence convinced scientists that sub-atomic particles exist?
6. How did Rutherford demonstrate that the positive charge in an atom is concentrated into a very small area?

PRACTICE QUESTIONS

1. Carbon-14 is a radioactive isotope of carbon. Which one of the following statements is true? [1 mark]

   A An atom of carbon-14 has 14 protons in its nucleus. ☐

   B The relative atomic mass of carbon is 14. ☐

   C An atom of carbon-14 has 6 electrons. ☐

   D An atom of carbon-14 contains the same number of protons as neutrons. ☐

2. Geiger and Marsden conducted experiments in which they fired α particles at gold foil. Some α particles were deflected backwards. The results of this experiment showed that [1 mark]

   A   the nucleus contains protons ☐

   B   the nucleus contains neutrons ☐

   C   there are the same number of protons and electrons ☐

   D   the nucleus is very small. ☐

3. Scientists now accept a model of the atom in which electrons in fixed energy orbitals exist in a relatively large space surrounding a very tiny nucleus in which most of the mass is found. Earlier models of atomic structure have now been rejected because [1 mark]

   A   earlier models are not based on scientific evidence ☐

   B   the current model provides a better explanation of observations ☐

   C   modern scientists are better able to share their ideas ☐

   D   modern scientific techniques are more accurate. ☐

4. Radioactive isotopes of americium are used as a component of smoke detectors. ^{241}Am is made in nuclear reactors and is a decay product of ^{242}Pu.

   a) How many protons are there in an atom of americium-241? [1 mark]

   b) What is meant by the fact that americium-241 is an isotope? [1 mark]

   c) In terms of sub-atomic particles, what is the difference between ^{241}Am and ^{242}Pu? [2 marks]

   d) When ^{241}Am decays, 2 protons and 2 neutrons are lost from the nucleus. Explain why the decay product is no longer americium. [2 marks]

5. Scientists working on the way food that is eaten is made into molecules in the body are able to 'label' food molecules with nitrogen-15.

   a) How many protons and neutrons are in an atom of nitrogen-15? [2 marks]

   b) How does the chemical behaviour of nitrogen-15 differ from that of nitrogen-14? [1 mark]

   c) How can scientists detect the difference between a molecule that contains nitrogen-14 and one that contains nitrogen-15? [1 mark]

   d) The A_r of nitrogen is 14.0. What does this tell us about the relative abundance of nitrogen-15? No calculation required. [1 mark]

6. Ideas about the structure of the atom have changed through the years. Describe the current model used by chemists and compare it to the billiard ball model. [4 marks]

DAY 1 — 60 Minutes

Representing Chemical Reactions

Interpreting Chemical Formulae

A chemical formula shows the ratio of each type of atom in a substance.

The subscript after each element symbol shows the number of atoms of that element in the formula.

Al_2O_3 $2 \times Al$ $3 \times O$ atoms

For substances containing more than one of the same compound ion, the formula for the compound ion is written in brackets with the number in subscript written outside the brackets.

$Mg(OH)_2$ $1 \times Mg$ $2 \times O$ $2 \times H$ atoms

$(NH_4)_2S$ $2 \times N$ $8 \times H$ $1 \times S$

$Al_2(SO_4)_3$ $2 \times Al$ $3 \times S$ $12 \times O$

No brackets are used if the formula contains only one compound ion, i.e. NaOH not Na(OH).

The formula never changes when balancing an equation.

Naming Chemicals from their Formula

Simple Binary Ionic Compounds, e.g. NaCl

The metal ion keeps its element name.	Na	sodium
The non-metal ion name ending changes to -ide.	Cl	chloride
The number of each type of atom does not affect the name.	Na_2S	sodium sulfide

Ionic Compounds Containing Transition Elements with Variable Oxidation Numbers

The charge on the ion is given in Roman numerals after the name.

$FeCl_3$	iron(III) chloride	$FeCl_2$	iron(II) chloride
CuO	copper(II) oxide	Cu_2O	copper(I) oxide

Naming Compound Ions from their Formula

General rules (with numerous exceptions!)

Write the first part of the first element name. Add the ending: -ate	CO_3^{2-}	**carbon**ate
	NO_3^-	**nitr**ate
	SO_4^{2-}	**sulf**ate
	PO_4^{3-}	**phosph**ate

Exceptions: OH^- hydroxide; NH_4^+ ammonium

Write the oxidation state of the non-oxygen element after the name.			
NO_2^-	nitrate(III)	NO_3^-	nitrate(V)
SO_4^{2-}	sulfate(VI)	SO_3^{2-}	sulfate(IV)
ClO^-	chlorate(I)	ClO_3^-	chlorate(V)
VO_2^+	vanadate(V)	VO^{2+}	vanadate(IV)

For an explanation of how to decide oxidation number see Day 4 Redox page 60.

Writing Chemical Formulae from Names

Each component of the formula can be considered to have a combining number.

Combining numbers of s- and p-block elements

Group	1	2	3	4	5	6	7
Combining number	1	2	3	4	3	2	1

For a compound ion the combining number is the same as the **number** of charges on the ion.

To write the formula, e.g. calcium chloride:

Write the symbol / formula of each part of the name.	Ca	Cl
Write the combining number for each.	2	1
Swap over the combining numbers.	1	2
These numbers become the subscript to the symbol / formula. (Do not write 1.)	Ca	Cl$_2$
Cancel to simplest ratio of atoms if required.	CaCl$_2$	

Further examples:

Magnesium sulfide		Potassium carbonate	
Mg	S	K	CO$_3$
2	2	1	2
Mg$_2$	S$_2$	2	1
MgS		K$_2$	CO$_3$
		K$_2$CO$_3$	

Writing Balanced Equations

Balanced equations describe a chemical change in terms of
- the formulae of the reacting substances
- the formulae of the products
- the quantities of reactants and products.

Since matter cannot be created or destroyed in a reaction there must be the same number of each type of atom in the reactants as in the products.

$$Ca(OH)_2 + CO_2 \longrightarrow CaCO_3 + H_2O$$

1×Ca 4×O 2×H 1×C ⟶ 1×Ca 4×O 2×H 1×C

To write a formula equation from a word equation:

1. Write the word equation.

iron(III) oxide + magnesium ⟶ magnesium oxide + iron

2. Write the formula of each reactant and product.

$$Fe_2O_3 + Mg \longrightarrow MgO + Fe$$

3. Count the number of atoms of each element in the reactants in turn and check it against the number of atoms of that element in the products.

$$Fe_2O_3 + Mg \longrightarrow MgO + Fe$$
Fe   2   1

4. If the numbers do not match for the first element, change the coefficients in front of the formulae until the number of atoms of the element in reactants and products is the same. **Do not change the formula**.

$$Fe_2O_3 + Mg \longrightarrow MgO + 2Fe$$

5. Check the next element and return to stage 3.

$$Fe_2O_3 + Mg \longrightarrow MgO + 2Fe$$
O   3   1

$$Fe_2O_3 + Mg \longrightarrow 3MgO + 2Fe$$
Mg   1   3

$$Fe_2O_3 + 3Mg \longrightarrow 3MgO + 2Fe$$

This process may have to be repeated several times for each element for more complex equations.

Example
Write a balanced equation from this word equation:

ammonia + oxygen ⟶ nitrogen(II) oxide* + water

ammonia + oxygen ⟶ nitrogen oxide + water

$$NH_3 + O_2 \longrightarrow NO + H_2O$$
N   1   1
H   3   2

Discrepancy so change coefficients, check N again

$$2NH_3 + O_2 \longrightarrow NO + 3H_2O$$
H   6   6
N   2   1

Discrepancy so change coefficients

$$2NH_3 + O_2 \longrightarrow 2NO + 3H_2O$$
N   2   2
O   2   2   3

9

DAY 1

Discrepancy so change coefficients

2NH₃ + **5**O₂ ⟶ **4**NO + **6**H₂O

O	10	4 6
N	2	4

Discrepancy so change coefficients

4NH₃ + 5O₂ ⟶ 4NO + 6H₂O

N	4	4
H	12	12
O	10	4 6

* See Redox on page 60 for how to work out the formula of compounds with variable oxidation states.

Chemical equations are written in terms of the **number** of particles, not the mass of particles.

When substances are weighed out in a laboratory, the chemical equation is important for calculating masses or moles of substances.

Ionic Equations

In an <u>ionic equation</u> any dissolved ions are written separately rather than as a compound.

$H_2SO_4(aq)$ is written as $2H^+(aq) + SO_4^{2-}(aq)$

Any ions that appear as both reactants and products in the equation are not included. (These are known as <u>spectator ions</u>.)

All substances that are not ions in solution are written as usual.

To write an ionic equation:

1. Write the full equation.
   $H_2SO_4(aq) + 2NaOH(aq) \longrightarrow Na_2SO_4(aq) + H_2O(l)$

2. Write all the aqueous ionic compounds as separate ions.
   $2H^+(aq) + SO_4^{2-}(aq) + 2Na^+(aq) + 2OH^-(aq) \longrightarrow 2Na^+(aq) + SO_4^{2-}(aq) + 2H_2O(l)$

3. Cancel any ions that appear on both sides of the equation.
   $2H^+(aq) + \cancel{SO_4^{2-}(aq)} + \cancel{2Na^+(aq)} + 2OH^-(aq) \longrightarrow \cancel{2Na^+(aq)} + \cancel{SO_4^{2-}(aq)} + 2H_2O(l)$

4. Cancel down to the simplest ratio of ions and what remains is the ionic equation.

5. $\cancel{2}H^+(aq) + \cancel{2}OH^-(aq) \longrightarrow \cancel{2}H_2O(l)$
   $H^+(aq) + OH^-(aq) \longrightarrow H_2O(l)$

It is essential to include state symbols in ionic equations.

Half Equations

Half equations can be used to describe redox reactions by separating out the oxidised and reduced reactants. They focus on the species that are 'doing' the chemistry.

The <u>reduction</u> half equation shows <u>electrons on the</u> reactants side. *Charge of product reduced*

The <u>oxidation</u> half equation has electrons on the products side.

Full equation
$Br_2(aq) + 2NaI(aq) \longrightarrow 2NaBr(aq) + I_2(aq)$

Ionic equation
$Br_2(aq) + 2I^-(aq) \longrightarrow 2Br^-(aq) + I_2(aq)$

Half equations
$Br_2(aq) + 2e^- \longrightarrow 2Br^-(aq)$ reduction
$2I^-(aq) \longrightarrow I_2(aq) + 2e^-$ oxidation

For more on how to decide what has been oxidised and what reduced and how to write half equations see Redox on page 60.

QUICK TEST

1. How many hydrogen atoms are there in the formula $(NH_4)_2CO_3$?

2. What is the formula of potassium sulfate?

3. Balance the equation:
   $H_3PO_4 + Mg \longrightarrow Mg_3(PO_4)_2 + H_2$

4. Write a balanced symbol equation for the reaction:
   magnesium nitrate + sodium carbonate ⟶ magnesium carbonate + sodium nitrate

5. Write an ionic equation for
   $2KI(aq) + Pb(NO_3)_2(aq) \longrightarrow 2KNO_3(aq) + PbI_2(s)$

6. Write the half equations for
   $Zn(s) + H_2SO_4(aq) \longrightarrow ZnSO_4(aq) + H_2(g)$

PRACTICE QUESTIONS

1. $(NH_4)_2SO_4$

This formula tells us that [1 mark]

A   it contains 4 hydrogen atoms; it is called ammonium sulfide ☐

B   it contains 4 hydrogen atoms; it is called ammonium sulfate ☐

C   it contains 8 hydrogen atoms; it is called ammonium sulfide ☐

D   it contains 8 hydrogen atoms; it is called ammonium sulfate. ☐

2. The correct name of KClO is [1 mark]

A   potassium chlorate(I) ☐   B   potassium chloride(I) ☐

C   potassium chlorate(II) ☐   D   potassium chloride(II). ☐

3. Sodium hydrogen carbonate is used as a raising agent in self-raising flour. When mixed with an acid it releases carbon dioxide. The carbon dioxide gas bubbles cause the cake to rise.

   a) Write the formula of sodium hydrogen carbonate. [1 mark]

   If the acid used was sulfuric acid then the overall reaction would be

   sodium hydrogen carbonate + sulfuric acid ⟶ sodium sulfate + carbon dioxide + water

   The formula for sulfuric acid is H_2SO_4.

   b) Write a balanced symbol equation for this reaction. [2 marks]

   Sulfuric acid is not suitable for use in self-raising flour but a variety of acids are used in different flours. All the reactions are essentially the same, however, since acids release H^+.

   c) Write an **ionic** equation for the reaction that occurs when a cake rises. Include state symbols. [3 marks]

4. Titanium is extracted from an ore known as rutile. This contains impure titanium(IV) oxide. The first step of extraction is heating the ore with chlorine and carbon to produce $TiCl_4$ and carbon monoxide. Next the $TiCl_4$ is heated with magnesium, which displaces the titanium metal and converts the magnesium to magnesium chloride.

   a) Write the formula of titanium(IV) oxide. [1 mark]

   b) Give the name of $TiCl_4$. [2 marks]

   c) Write a balanced symbol equation for the first step of the extraction. [2 marks]

   In the second step of the extraction magnesium is oxidised to form magnesium ions.

   d) Give the formula of a magnesium ion. [1 mark]

   e) Write a half equation for the reaction of magnesium in this step of the reaction. [1 mark]

   Some of the $TiCl_4$ produced at the end of the first step can be reacted with hydrogen to produce HCl(g) and $TiCl_3$(s), which is used as a catalyst in the polymer industry.

   f) Write a word equation for this reaction. [2 marks]

DAY 1 — 60 Minutes

Electron Configuration

Electron Shells

Electrons exist in shells of different energies around the nucleus.

Each shell is labelled with a **principle quantum number**, n. *basically orbital #*

The higher the principle quantum number, the further from the nucleus the shell and the higher the energy of the electrons. Each subsequent shell can hold more electrons.

Shell number	Principle quantum number	Maximum number of electrons
1	$n = 1$	2
2	$n = 2$	8
3	$n = 3$	18
4	$n = 4$	32

Sub-shells and Orbitals

An orbital is a region around the nucleus where an electron is likely to be found. It is found by mathematical calculation. Two electrons can exist in the same orbital provided they have **opposite spin**.

Electron shells are divided into sub-shells. The higher the principle quantum number of the shell, the more sub-shells it holds.

There are four different types of sub-shell, s, p, d and f, each with different numbers of orbitals.

s sub-shells hold 1 orbital and have a spherical shape.

p sub-shells hold 3 orbitals and have a dumb-bell shape.

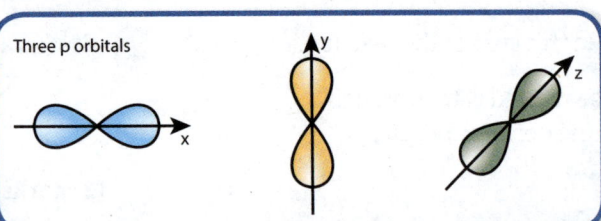

Three p orbitals

d sub-shells hold 5 orbitals.

Sub-shells in each shell:

Shell	Sub-shells
1	s
2	s p
3	s p d
4	s p d f

Electrons in Boxes

Electrons in boxes is a convention for showing electron structure. One box represents one orbital.

Electrons enter the lowest energy orbital available. They fill one electron per orbital for each sub-shell before a second electron enters the orbital, e.g. iron:

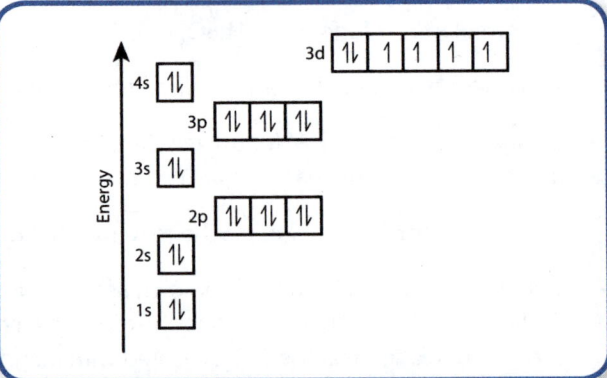

The 4s sub-shell is of lower energy than the 3d and is filled before the 3d begins filling.

This can also be shown with all boxes inline, e.g. oxygen:

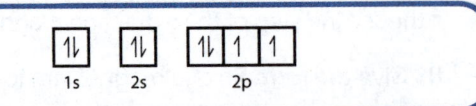

Anomalous - deviating from what is normal

Electron Configuration

The electron configuration of an element is the number of electrons in each shell and sub-shell. It is written using the convention shown below.

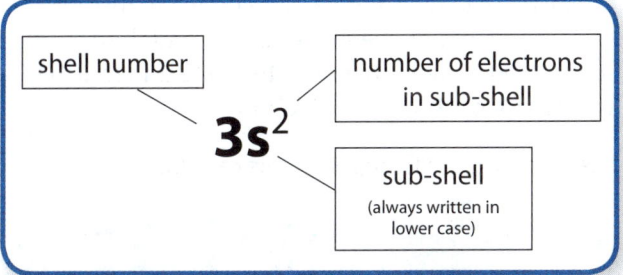

e.g. The electronic structure of Fe is

[Ne]

$1s^2\ 2s^2\ 2p^6\ 3s^2\ 3p^6\ 3d^6\ 4s^2$ ← *this fills first*

2   8   *These are the remaining*

All the third shell sub-shells are written together even though the filling order is 4s then 3d.

Electronic configuration can be abbreviated by using the symbol for a noble gas to represent the electronic configuration for that gas and then listing the remaining electrons.

[Ne] is an abbreviation for $1s^2\ 2s^2\ 2p^6$

[Ar] is an abbreviation for $1s^2\ 2s^2\ 2p^6\ 3s^2\ 3p^6$

e.g. chlorine [Ne] $3s^2\ 3p^5$ iron [Ar] $3d^6\ 4s^2$

Anomalous electron configurations include

copper Cu [Ar] $3d^{10}\ 4s^1$ **not** [Ar] $3d^9\ 4s^2$ and
chromium Cr [Ar] $3d^5\ 4s^1$ **not** [Ar] $3d^4\ 4s^2$

Electron Configurations of Ions

When a positive ion is formed the electrons are removed in the reverse order to that in which they are written.

Calcium Ca ⟶ Ca^{2+} [Ar] $4s^2$ ⟶ [Ar]
Iron   Fe ⟶ Fe^{2+} [Ar] $3d^6\ 4s^2$ ⟶ [Ar] $3d^6$

This means the 4s electrons are removed before the 3d electrons and that electrons are not always removed in the reverse of their filling order. The reason for this is that these are further from the nucleus than the 3d electrons since they are in the outer shell.

For s- and p-block elements, the electronic configuration of all positive ions in the same period is the same, since they are all losing electrons to achieve a full outer shell. The same is true of all negative ions in the same period that are gaining electrons to achieve a full outer shell.

The electronic configuration of negative ions from one period is the same as those of positive ions in the next period, e.g.

N^{3-}	$1s^2\ 2s^2\ 2p^6$
O^{2-}	$1s^2\ 2s^2\ 2p^6$
F^-	$1s^2\ 2s^2\ 2p^6$
Na^+	$1s^2\ 2s^2\ 2p^6$
Mg^{2+}	$1s^2\ 2s^2\ 2p^6$
Al^{3+}	$1s^2\ 2s^2\ 2p^6$

Evidence for Electron Sub-shells

The first ionisation energy of an element is the enthalpy change when one mole of electrons is removed from one mole of atoms in their gaseous state.

$$X(g) \longrightarrow X^+(g) + e^-$$

The first ionisation energies of elements across a period:

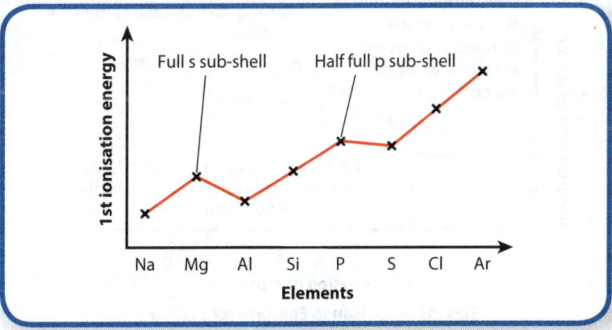

★ The general trend is that the first ionisation energy of elements increases across a period. Each subsequent element has one additional proton in the nucleus but similar shielding from electron shells, and the atoms are smaller so there is a greater attraction of outer shell electrons to the nucleus.

lower 1st ionisation energy → higher period

DAY 1

The detailed changes in first ionisation energy provide evidence for the existence of sub-shells and orbitals. For example, for Period 3 there is a drop in first ionisation energy between Mg and Al. This reflects the fact that the first ionisation energy of Al requires an electron to be removed from a p sub-shell. p sub-shells are at a higher energy than s sub-shells, so less energy is needed to remove an electron.

There is a second drop in first ionisation energy between phosphorus and sulfur. This is because the electron for phosphorus is unpaired and removed from the p orbital. But the sulfur is losing an electron from a doubly occupied orbital. Repulsion between electrons as they pair up reduces the energy needed to remove them.

Successive Ionisation Enthalpies

The second ionisation enthalpy of an element is the enthalpy change when one mole of electrons is removed from one mole of unipositive (1+) ions in their gaseous state.

$$X^+(g) \longrightarrow X^{2+}(g) + e^-$$

2nd e⁻ removed from each atom

The third ionisation enthalpy is removing an electron from the 2+ ion, the fourth from a 3+ ion and so on.

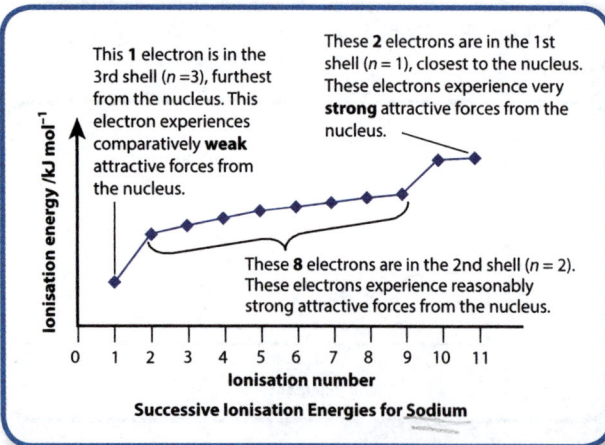

Successive Ionisation Energies for Sodium

Each successive ionisation energy increases because it requires more energy to remove a negative electron from an increasingly positive ion. There is a large jump in ionisation energy when an electron is being removed from the next shell in.

Evidence from ionisation energies can be used to identify the group of an element.

E.g. Successive molar ionisation energies of an element in kJ mol⁻¹ are: G3?

800.6	2427.1	3659.7	25 025.8	32 826.7

The increase in ionisation energy between the first and second and between the second and third is about 1500 kJ mol⁻¹. Between the third and fourth is an increase of about 21 000 kJ mol⁻¹.

The element is in Group 3 (in this case boron).

QUICK TEST

1. State how many orbitals a p sub-shell holds.
2. What is the principle quantum number of the highest electron shell for aluminium?
3. Sketch a p-orbital.
4. Draw the electron in boxes configuration for nitrogen.
5. Write the electron configuration for a chlorine atom.
6. Name the element with the electron configuration: $1s^2\ 2s^2\ 2p^6\ 3s^2\ 3p^6\ 3d^3\ 4s^2$

PRACTICE QUESTIONS

1. The electronic configuration of zinc is [1 mark]

   A [Ar] $3d^{10}$ $4s^2$

   B $1s^2$ $2s^2$ $2p^6$ $3s^2$ $3p^6$ $3d^{10}$

   C [Ar] $1s^2$ $2s^2$ $2p^6$ $3s^2$ $3p^6$ $3d^{10}$ $4s^2$

   D $1s^2$ $2s^2$ $2p^6$ $3s^2$ $3p^6$ $3d^{10}$ $4s^2$ $4p^6$

2. The electron arrangement of a carbon atom may be represented as [1 mark]

   A [↑↓] [↑↓] [↑] [↑] [↑]

   B [↑↓] [↑↓] [↑] [↑]

   C [↑↓] [↑↓] [↑↓]

   D [↑↓] [↑] [↑↓] [↑]

3. The successive ionisation energies of an element are shown below.

Ionisation energy kJ mol^{-1}	1st	2nd	3rd	4th	5th	6th	7th
A	590	1145	6491	8153	10 496	12 270	14 206
B	786	1577	3228	4354	16 100	19 805	23 780
C	1314	3388	5300	7469	10 989	13 326	71 330

   a) Write an equation that describes the 2nd ionisation energy of carbon. [2 marks]

   b) Why does the ionisation energy increase with each successive electron removal? [2 marks]

   c) To which group does element **B** belong? [1 mark]

   d) Predict the value of the 8th ionisation energy of element **C**. [1 mark]

   e) The first ionisation energy of magnesium is 738 kJ mol^{-1}. Use this information and your knowledge of the general trends in ionisation energies to suggest the identity of element **A**. Justify your answer. [4 marks]

4. a) Describe and explain the general trend in first ionisation energies of elements across a period. [2 marks]

   b) Explain how the change in first ionisation energy of elements across Period 2 provides evidence for the existence of electron sub-shells. [4 marks]

DAY 1 — 60 Minutes

Calculations 1

A **mole** is the unit for the amount of a substance.

1 mole = 6.02×10^{23} particles of that substance.
Particles can refer to anything: atoms, molecules, ions, electrons.

The molar mass is the mass of 6.02×10^{23} particles of that substance measured in g mol^{-1}.

Molecular Mass and Relative Formula Mass

The term relative molecular mass is used for substances that consist of molecules, simple covalent substances. It is the mean mass of a molecule of a substance relative to $\frac{1}{12}$ mass of an atom of carbon-12.

The relative formula mass, RFM, is used for substances that do not have molecules, such as ionic or giant covalent substances. It is the mean mass of a formula unit of the substance relative to $\frac{1}{12}$ mass of an atom of carbon-12.

The symbol M_r may be used to represent either molecular mass or relative formula mass.

To calculate M_r or RFM, add together the A_r of each atom in the substance, e.g. RFM of ammonium sulfate $(NH_4)_2SO_4$

$= (2 \times N) + (8 \times H) + (1 \times S) + (4 \times O)$
$= (2 \times 14) + (8 \times 1) + (1 \times 32) + (4 \times 16) = 132$

Amounts of Substance

Calculating Reacting Masses

Use the reaction equation to calculate masses of one substance in the reaction equation when the mass of the other substance is given, e.g.

What mass of iron would be made from 100 g of iron oxide in the following reaction?

$$Fe_2O_3 + 3CO \longrightarrow 3CO_2 + 2Fe$$

1. Find the M_r of the two substances in the question.

   What mass of iron would be made from 100 g of iron oxide?

   $Fe_2O_3 = 159.6$ $Fe = 55.8$

2. Calculate the moles of the substance for which the mass is given. [**moles = mass/M$_r$**]

   What mass of iron would be made from 100 g of iron oxide?

   $\frac{100}{159.6} = 0.627$ moles Fe_2O_3

3. Find the mole ratio of the known substance to the unknown substance. The mole ratio is given by coefficients in the equation.

   $$1Fe_2O_3 + 3CO \longrightarrow 3CO_2 + 2Fe$$

   Moles of Fe_2O_3 are known.

   1 Fe_2O_3 : 2Fe   =   1 : 2

4. Multiply moles of the known substance by the mole ratio to find the moles of the second substance.

   Moles $Fe_2O_3 = 0.627$

   Moles Fe $= 0.627 \times 2 = 1.254$

5. Calculate the mass of the second substance.

   Mass = $M_r \times$ moles

   $1.254 \times 55.8 = 69.9732 = 70.0$ g Fe

Finding the Moles of Water of Crystallisation

Use the mass of hydrated crystals before and after heating, e.g.

What is the value of X in a sample of $CuSO_4.XH_2O$ if the mass before heating is 30.00 g and after heating to constant mass is 19.20 g?

1. Find the mass of water driven off during heating by subtracting the final mass of anhydrous crystals from the initial mass of hydrated crystals.

   $$CuSO_4.XH_2O \longrightarrow CuSO_4 + XH_2O$$

   Mass before heating is 30.00 g and after heating to constant mass is 19.20 g.

   $30.00 - 19.20 = 10.8$ g H_2O driven off

2. Find the moles of water driven off.

Moles = $\frac{\text{mass of H}_2\text{O}}{M_r \text{H}_2\text{O}}$

$M_r \text{H}_2\text{O} = 18$ Moles $\text{H}_2\text{O} = \frac{10.8}{18} = 0.6$ moles H_2O

3. Find the moles of anhydrous crystals from the mass after heating.

…after heating to constant mass is (19.20 g)

$M_r \text{CuSO}_4 = 159.6$
Moles = $\frac{19.2}{159.6} = 0.12$ moles of CuSO_4

4. Find the mole ratio of the anhydrous crystals to water to find the value of X.

Moles anhydrous $\text{CuSO}_4 : \text{H}_2\text{O}$
$0.12 : 0.6 = 1 : 5$ X = 5
Formula = $\text{CuSO}_4 \cdot 5\text{H}_2\text{O}$

Finding Concentrations by Titration

Use the reaction equation to calculate the concentration of one reactant when the concentration and volume of the other reactant are given, e.g.

22.70 cm³ of 0.01 mol dm⁻³ Na_2CO_3(aq) were needed to neutralise 25 cm³ HCl(aq). What is the concentration of HCl(aq)?

$2\text{HCl} + \text{Na}_2\text{CO}_3 \longrightarrow 2\text{NaCl} + \text{H}_2\text{O} + \text{CO}_2$

1. Find the moles of the reactant for which concentration and volume are given.

Moles = conc × volume in dm³

(22.70 cm³) of (0.01 mol dm⁻³) Na_2CO_3(aq) were needed to neutralise 25 cm³ HCl(aq).

1 dm³ = 1000 cm³   Convert from cm³ to dm³ by dividing by 1000.

Moles $\text{Na}_2\text{CO}_3 = 0.01 \times 0.0227 = 0.000227$ moles

2. Calculate the moles **in 25 cm³** of the other reactant using the mole ratio.

(2)HCl + (1)$\text{Na}_2\text{CO}_3 \longrightarrow 2\text{NaCl} + \text{H}_2\text{O} + \text{CO}_2$

Mole ratio 1Na_2CO_3 : 2HCl = 1 : 2
Moles HCl in 25 cm³ = 2 × 0.000227
= 0.000454

3. Calculate the concentration of the second reactant.

Concentration = moles/volume in dm³

Conc HCl = $\frac{0.000454}{0.025} = 0.0182$ mol dm⁻³

Changing the units of concentration from mol dm⁻³ to g dm⁻³   *These can't both be true?*

mol dm⁻³ × M_r = g dm⁻³ | g dm⁻³ = $\frac{\text{mol dm}^{-3}}{M_r}$

e.g. M_r HCl = 36.5
0.1 mol dm⁻¹ HCl = 0.1 × 36.5 = 3.65 g dm⁻³

M_r NaOH = 40
5 g dm⁻³ NaOH = $\frac{5}{40} = 0.125$ mol dm⁻³

Calculating the Purity of a Sample

Use titration results to calculate the % by mass of a component in a sample, e.g. 8 g of an impure sample of NaOH was dissolved in 250 cm³ water. 25 cm³ of this solution required 37.50 cm³ of 0.4 mol dm⁻³ HCl to reach the end point in a titration. What was the percentage purity of the sample?

1. Calculate the moles of NaOH in 25 cm³ of the sample.

$0.4 \times 0.0375 = 0.015$ mol HCl in 37.5 cm³
Mole ratio 1 : 1
0.015 mol NaOH in 25 cm³

2. Calculate the moles of NaOH in the whole sample. 8 g of an impure sample of NaOH was dissolved in (250 cm³) water.

1 cm³ contains $\frac{0.015}{25}$ mol NaOH
250 cm³ contains $\frac{0.015}{25} \times 250 = 0.15$ mol NaOH

3. Calculate mass of NaOH
Mass = M_r × moles
M_r NaOH = 40
Total mass in sample = 0.15 × 40 = 6 g

4. Calculate % of the total mass of the sample that was the compound.

% by mass = $\frac{\text{mass of compound}}{\text{total mass}} \times 100\%$

(8 g) of an impure sample of NaOH was dissolved in 250 cm³ water.

$\frac{6}{8} \times 100 = 75\%$ of NaOH in total sample

DAY 1

solvent = substance in which solute is dissolved

Diluting Solutions

For a solution:

moles = concentration (mol dm^{-3}) × volume (dm^3)

When a solution is diluted, only the volume of solvent changes, not the number of moles of solute. The concentration × volume of the concentrated solution is the same as the concentration × volume of the diluted solution.

$$C_{(original)} \times V_{(original)} = C_{(new)} \times V_{(new)}$$

So $V_{(new)} = \dfrac{C_{(original)} \times V_{(original)}}{C_{(new)}}$

To dilute a solution to a known concentration:

1. Remove a known volume of known concentration solution.
2. Calculate the volume of the new solution using
$$V_{(new)} = \dfrac{C_{(original)} \times V_{(original)}}{C_{(new)}}$$
3. Add solvent to make the total volume of solution up to the volume calculated.

To find the new concentration of a solution to which solvent has been added:

$$C_{(new)} = \dfrac{C_{(original)} \times V_{(original)}}{V_{(new)}}$$

> **Example**
> What is the new concentration of 25 cm^3 of 0.1 mol dm^{-3} NaOH that has been diluted to 100 cm^3?
>
> $C_{(new)} = \dfrac{0.1 \times 25}{100} = 0.025$ mol dm^{-3}

Significant Figures

Significant figures are counted from the first digit of the number that is not zero (left to right).

Number	\multicolumn{4}{c	}{Significant figures}
	1st	2nd	3rd	4th
4620	4	6	2	0
0.0058	5	8	No further sig figs	

Rounding to a Specified Number of Significant Figures

E.g. Write these numbers to 3 significant figures.

	54782	0.0692433
Count the significant figures required.	547̂82	0.0692̂433
Check the digit to the right of the last significant figure.	5478̂2	0.06924̂33
If it is 4 or lower write all the figures to the right of the last figure as 0.		0.0692000
If it is 5 or higher add 1 to the last significant figure and write all other digits as 0.	547̂82 → 548̂00	
For numbers smaller than 0 remove all zeros after the last significant figure.		0.0692
Number to 3 s.f.	54800	0.0692

- A precise measurement has a small random error. Repeated precise measurements should give values that are very close together.
- The number of significant figures shows the precision of the number.
- The number of significant figures quoted in an answer should not be higher than the lowest number of significant figures in the data used for the calculation.

QUICK TEST

1. Calculate the M_r of $CaCl_2.6H_2O$.
2. What mass of CaO would be made from the thermal decomposition of 10 g of $CaCO_3$?
3. How many moles of NaOH are in 50 cm^3 of a 0.125 mol dm^{-1} solution?
4. How many moles of $CuCO_3$ are in 12.35 g?
5. Give the mole ratio of H_2 : NH_3 in
   $3H_2 + N_2 \rightleftharpoons 2NH_3$
6. What is the concentration of a 0.1 mol dm^{-3} solution of NaOH in g dm^{-3}?

18

PRACTICE QUESTIONS

1. 25 cm³ of a solution of HCl reacts with exactly with 50 cm of 0.1 mol dm⁻³ NaOH. The concentration of HCl is [1 mark]

   A 0.1 mol dm⁻³ ☐   B 7.3 g dm⁻³ ☐

   C 0.05 mol dm⁻³ ☐   D 0.0073 g dm⁻³ ☐

2. 150 g of CaCO₃ was strongly heated to constant mass.

   CaCO₃ ⟶ CaO + CO₂

   The final mass would be [1 mark]

   A 84.07 g ☐   B 100.01 g ☐

   C 66.00 g ☐   D 42.08 g ☐

3. A scientist was trying to discover the value of X in CaCl₂.XH₂O. She weighed a sample of CaCl₂.XH₂O in a crucible and then heated it repeatedly in a furnace. The following results were obtained.

Starting mass of crucible (g)		20.05
Mass of crucible + sample (g) Before heating		23.78
After heating	1	22.43
	2	22.06
	3	21.94
	4	21.94

   a) Why was the crucible heated four times? [1 marks]
   b) Why was it not necessary to heat the crucible more than four times? [2 marks]
   c) Use the results to calculate the value of X. Show all your working. [5 marks]

4. Aluminium oxide can be reduced to aluminium by electrolysis.

   $$2Al_2O_3 \longrightarrow 4Al + 3O_2$$

   A typical industrial aluminium smelter contains a large number of electrolysis cells, each of which produces about 1 tonne of aluminium a day. The oxygen produced in the electrolysis reacts with the carbon (graphite) anode to produce carbon dioxide. This means that the graphite anodes have to be continually replaced.

   a) What mass of aluminium oxide would be required to produce 1 tonne of aluminium? [2 marks]
   b) Assuming that the graphite electrodes are pure carbon and assuming that all of the oxygen produced at the anode reacts to produce carbon dioxide

   C + O₂ ⟶ CO₂

   what mass of graphite is lost from the electrode in production of 1 tonne aluminium? [3 marks]
   c) In reality, some of the oxygen forms carbon monoxide rather than carbon dioxide. Explain if this means that the graphite anodes have to be removed more or less frequently than expected. [1 mark]

5. A sample of baking soda was labelled as containing 95% NaHCO₃. 2.0 g of the baking soda was dissolved in water and made up to 250 cm³. A 25 cm³ sample of this was titrated against HCl and required 20.05 cm³ of 0.1 mol dm⁻³ HCl to reach an end point. Was the claim on the label correct? Show your working. [4 marks]

DAY 2 — 60 Minutes

Ionic Bonding and Structure

what do these blocks mean?

The Meaning of Bonding and Structure

Bonding is either ionic, covalent, or metallic.

Structure can be giant (also called lattice, crystal, or network) or simple (also called molecular).

The bonding and structure of a substance determine its physical properties. The physical properties can be explained using bonding and structure.

> **Don't forget**
>
> … if a question asks you about structure and bonding you must include a description of both.

Ionic Bonding

Ionic bonding is **electrostatic attraction** between oppositely charged ions, e.g.

$$[Mg]^{2+} \; 2[{\times\!\!\times\atop\times\!\!\times}Cl{\times\!\!\times\atop\times\!\!\times}]^-$$

Ions are formed when electrons are transferred from one species to another in order to give a more stable electronic configuration.

Monoatomic ions are atoms that have lost or gained electrons.

Metals lose their outer shell electrons to form positive ions (cations). Non-metals gain electrons to form negative ions (anions).

Elements in the main groups of the periodic table form ions in order to have a noble gas electron configuration. All elements in the same group have the same number of electrons in their outer shell, so form ions with the same charge.

Group	1	2	3	4	5	6	7	0
Charge on ion	1+	2+	3+	*	3–	2–	1–	*

*Does not form ions.

Many d-block elements can form more than one stable ion, e.g. Fe^{2+}, Fe^{3+}, Cu^+, Cu^{2+}.

Some form only one ion, e.g. Ag^+, Zn^{2+}.

Isoelectronic species have the same electron structure, e.g. N^{3-}, O^{2-}, F^-, Ne, Na^+, Mg^{2+}, Al^{3+}; all have the electronic configuration $1s^2\ 2s^2\ 2p^6$.

Compound ions are compounds that have lost or gained electrons.

The names and formulae of some important compound ions are:

NO_3^- nitrate(V)	HCO_3^- hydrogen carbonate
CO_3^{2-} carbonate	SO_4^{2-} sulfate
NH_4^+ ammonium	OH^- hydroxide

Formulae of Ionic Compounds

Ionic compounds are neutral so the charges on the positive ions equal the charges on the negative ions. Multiplying the number of positive and negative ions to achieve a neutral compound gives the ratio of ions. E.g. Given the formula of these ions:

Na^+ Mg^{2+} Cl^- OH^- S^{2-} CO_3^{2-}

the formulae of these compounds are

NaCl $MgCl_2$ NaOH $Mg(OH)_2$

Na_2S MgS Na_2CO_3 $MgCO_3$

Ionic Structures

Oppositely charged ions are held together in a **giant** 3D lattice by strong electrostatic attraction, e.g.

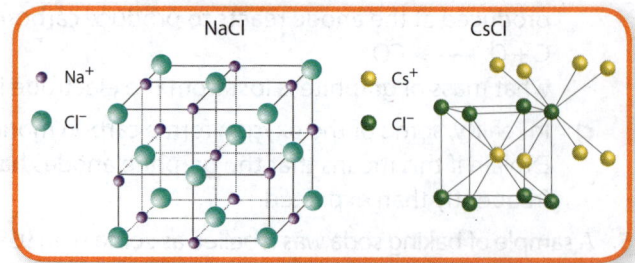

20

Properties

Ionic lattices have a high melting point: A large amount of energy is needed to break the strong electrostatic attraction between ions.

Ionic solids are non-conductors of electricity: There are no charged particles that can move through the lattice to carry the charge. All electrons are located within their ion and the ions are in a fixed position.

Molten or dissolved ionic substances are conductors: The ions can move and carry the charge between electrodes. Ions are discharged and become atoms when they reach the electrode, e.g.

$Na^+ + e^- \longrightarrow Na$ $2Cl^- \longrightarrow Cl_2 + 2e^-$

Many ionic substances are water soluble: The charged ions can become surrounded by polar water molecules (see page 25).

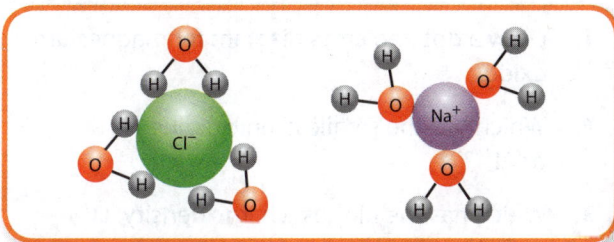

Not all ionic substances are soluble, e.g. $CaCO_3$, AgCl, $BaSO_4$ are all insoluble.

Ionic Radii

Cations have a smaller radius than their atoms because they have lost their outer shell electrons. The ionic radius of cations decreases across a period because there is one extra proton pulling on the same number of electrons with similar shielding.

Atomic radius: $Na^+ > Mg^{2+} > Al^{3+}$

Anions have a greater radius than their atoms because they have additional electrons in their outer shell. The ionic radius of anions decreases across the period because there is one fewer additional electron with each element.

Atomic radius: $P^{3-} > S^{2-} > Cl^-$

The radius of both cations and anions increases going down a group because the number of shells of electrons increases.

Charge Density

Charge density of ions is $\frac{\text{charge on ion}}{\text{ionic radius}}$

Charge density decreases down a group because the ionic radius increases and charge stays the same.

The charge density of **metal** ions increases across a period. The charge on the ions increases and the radius decreases.

The charge density of non-metal anions decreases across a period because the charge on the ion decreases by one for each element but the radius decreases only slightly.

The radius of **non-metal** ions is larger than that of metal ions in the same period because they have an extra shell of electrons. In an ionic compound the radius of the anion is almost always larger than the radius of the cation. (Exceptions are compounds of Rb, Cs and Ba with F.)

As the charge density of an ion increases, the strength of ionic bonding increases. This means an increase in melting point, e.g.

Melting point of

Group 2 salt > Group 1 salt $MgCl_2$ > NaCl

Period 2 salt > Period 3 salt   LiCl > NaCl > KCl

NaF > NaCl > NaBr

Distortion of Anion Shape

When the charge density of the cation is very high it distorts the shape of the anion.

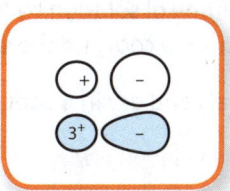

This gives the ionic bond some covalent nature.

The carbonates and nitrates of Group 2 become more resistant to thermal decomposition going down the group. This is because the higher charge density of metals at the top of the group distorts the shape of the carbonate or nitrate ion. This distortion makes the carbonate and nitrate ions easier to decompose.

DAY 2

Solubility Rules of Ionic Compounds

All sodium, potassium and ammonium compounds are soluble.

All nitrates and ethanoates are soluble.

Most chlorides are soluble. Exceptions: AgCl, $PbCl_2$.

Most sulfates are soluble. Exceptions: $BaSO_4$, $PbSO_4$, $CaSO_4$.

Most metal oxides, hydroxides, carbonates and phosphates are insoluble. Exceptions: sodium, potassium and ammonium compounds.

Preparing Salts

Selecting Reactants

Salts are made by the reaction of an acid (a chemical that releases hydrogen ions into solution) and a base (a chemical that reacts with an acid). The positive ion in the salt comes from the base and the negative ion from the acid.

Soluble Salts from an Insoluble Base

Excess insoluble base is added to acid. Unreacted base is removed by filtration.

1. Warm the appropriate acid, e.g. HCl for chlorides, HNO_3 for nitrates, H_2SO_4 for sulfates.
2. Add the appropriate insoluble base (the oxide or hydroxide of the correct metal) until no more reacts.
3. Filter to remove the excess base.
4. Reduce the volume of solution by heating, and then leave the solution to cool and the salt to crystallise.
5. Filter the crystals and dry in a cool oven.

Insoluble Salts by Precipitation

1. Choose **soluble** salts of the two ions in the required insoluble salt, e.g. a nitrate of the metal ion and a sodium salt of the negative ion.
2. Mix the two solutions in an appropriate ratio to precipitate the salt.
3. Filter, then wash and dry the precipitate.

Soluble Salts by Titration

Where no insoluble base is available, the pure salt is made by mixing exact quantities of the soluble ions. The exact quantities are determined by titration.

1. Fill a burette with the appropriate acid.
2. Titrate against a known volume of appropriate alkali using an indicator and note the volume required to reach the endpoint.
3. Repeat the titration without indicator.
4. Reduce the volume by heating and leave to crystallise.
5. Filter and dry the crystals in a cool oven.

QUICK TEST

1. Draw a dot and cross diagram for magnesium oxide.
2. Which has the smallest ionic radius: Na^+ or Al^{3+}?
3. Which has the highest charge density: Li^+ or Rb^+?
4. Which substance would you expect to have the highest melting point: KCl or $MgCl_2$?
5. Give the formula of a metal ion that is isoelectronic with O^{2-}.
6. Describe what happens when electricity is passed through a solution of ions.

PRACTICE QUESTIONS

1. Which compound is not ionic? **[1 mark]**

   A Potassium iodide   B Aluminium oxide
   C Ammonium sulfate   D Chloromethane

2. Which compound forms the strongest ionic bonds? **[1 mark]**

   A MgS   B MgCl$_2$
   C NaCl   D Na$_2$S

3. Which statement is true? **[1 mark]**

   A Sodium has a higher melting point than potassium because it donates more electrons to the delocalised sea of electrons.

   B Sodium has a lower melting point than magnesium because it has a lower charge density.

   C Magnesium has a higher melting point than potassium because it has a larger ionic radius.

   D Potassium has a higher melting point than sodium because it has a higher charge density.

4. Solid anhydrous calcium chloride is used as a drying agent in the preparation of organic liquids.

   a) Draw a dot and cross diagram for calcium chloride. **[4 marks]**

   b) Explain why solid calcium chloride does not melt even if used with hot organic liquids. **[2 marks]**

   c) Common table salt is sodium chloride. Describe the structure of sodium chloride and draw a labelled diagram showing the structure. **[3 marks]**

   d) Discuss the reasons for the difference in melting point between potassium chloride and calcium chloride. **[3 marks]**

   e) Both sodium chloride and sodium are solids at room temperature. Sodium chloride does not conduct electricity as a solid but sodium does. Use ideas about structure and bonding to explain this difference. **[6 marks]**

5. Lead(II) chromate is a yellow insoluble salt that was used historically as yellow pigment for paint. Chromate is a compound ion with the formula CrO$_4^{2-}$. Lead chromate occurs naturally in the mineral crocoite but can be made by ionic precipitation.

   a) Write the formula for lead(II) chromate. **[1 mark]**

   b) Suggest reagents for the production of lead(II) chromate by ionic precipitation and write an ionic equation for the reaction. **[3 marks]**

   c) Outline a method to obtain a dry sample of lead(II) chromate. **[2 marks]**

23

DAY 2   60 Minutes

Covalent Bonding and Structure

A covalent bond is a shared pair of electrons. The nuclei of two different atoms are electrostatically attracted to the same pair of electrons.
E.g. H_2

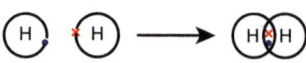

This gives both atoms a more stable electron configuration. For many elements this results in each atom having 8 electrons in its outer shell (a stable octet), e.g.

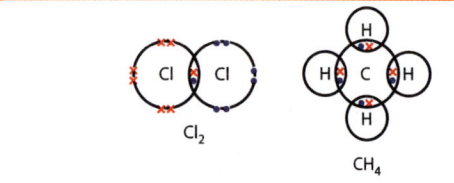

Covalent bonds usually form between two non-metals but there are exceptions, e.g. aluminium chloride.

How Many Covalent Bonds Can an Element Form?

Electrons in the outer shell that are part of a covalent bond are called **bonding electrons**; those that are not part of a bond are called **lone pairs**. In general:

Group number of element	Number of covalent bonds it forms	Number of lone pairs
4	4	0
5	3	1
6	2	2
7	1	3

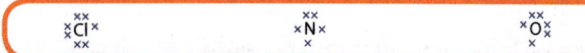

Molecules with an Expanded Octet

Elements in Period 2 have their electrons in s and p orbitals. There is no space for any more orbitals.

Elements in Period 3 and higher have more space for electrons in their outer shell. They have 'empty' d orbitals, and they can separate lone pairs of electrons so that each electron is alone in an orbital and can form a covalent bond.

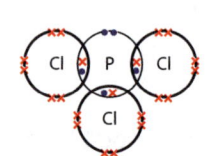 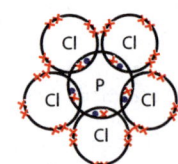

PCl_3: only 3 electrons from the outer bonding shell of phosphorus are used in bonding.

PCl_5: all 5 electrons from the outer bonding shell of phosphorus are used in covalent bonds.

Dative (Coordinate) Covalent Bonds

A covalent bond can form between two atoms where both of the shared electrons come from only one of the atoms.
E.g. NH_4^+

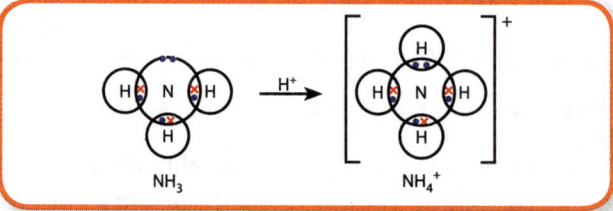

The direction of the donation of an electron in a dative covalent bond can be shown with an arrow from the donating atom to the receiving atom, e.g.

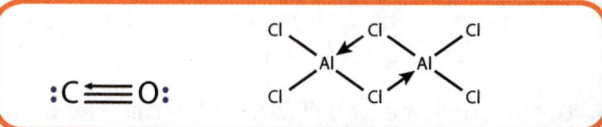

Note: $AlCl_3$ forms the dimer Al_2Cl_3 when it vaporises, which breaks back down to $AlCl_3$ as the temperature increases.

Multiple Bonds

Some elements are able to share more than one pair of electrons to form double or triple bonds.

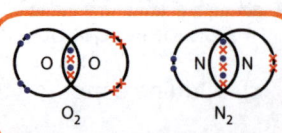

Do the electrons spin around? Or are they stationary?

In a single covalent bond the shared electrons can be found between the nuclei of the two atoms in a **sigma (σ) bond**.

In double bonds one pair of electrons are in a σ bond and the second shared pair of electrons can be found above and below the sigma bond in a **pi (π) bond**. E.g. ethene

A triple bond contains one σ bond and two π bonds at right angles to each other, e.g. for ethyne

A single sigma bond allows the atoms at either side to rotate freely. The π bond prevents the two atoms rotating so groups on either side of a π bond are fixed in position relative to each other.

Strength of Covalent Bonds

Bond length is the distance between the two nuclei sharing the electrons. Bond length increases with increasing atomic radius.

The longer the bond, the weaker the bond. C—C bonds are stronger than Si—Si bonds because of the smaller radius of the carbon atom.

π bonds are weaker than σ bonds. A double bond is less than twice as strong as a single bond. *half*

π bonds are likely to be attacked by electrophiles (species attracted to a negative charge) because the delocalisation of the electrons makes them easier to access.

Polar Covalent Bonds

Electronegativity of elements is the ability to attract electron density in a covalent bond. Fluorine is the most electronegative element, followed by oxygen. Noble gases do not readily form covalent bonds and so do not have an electronegativity value.

Electronegativity increases as the radius of the atom decreases and the number of protons pulling on the same shell of electrons increases. This means there is an increase in electronegativity across a period and up a group of the periodic table.

Each element can be given an electronegativity value using the Pauling scale.

H 2.2						
Li 1.0	Be 1.6	B 1.8	C 2.5	N 3.0	O 3.4	F 4.0
Na 0.9	Mg 1.3	Al 1.6	Si 1.9	P 2.2	S 2.6	Cl 3.2
K 0.8						Br 3.0
						I 2.7

Polar covalent bonds form where there is a significant difference in electronegativity between the two atoms sharing the electrons. This results in an uneven distribution of electrons between the two atoms sharing the electrons. The more electronegative atom has a greater share of the electrons and becomes slightly negative (δ−) while the other atom becomes slightly positive (δ+), e.g.

$^{δ+}$C=O$^{δ-}$ $^{δ+}$C—Cl$^{δ-}$ $^{δ-}$S—H$^{δ+}$ $^{δ+}$H—O$^{δ-}$ $^{δ+}$H—N$^{δ-}$

The greater the difference in electronegativity the more polar the bond. Very polar covalent bonds are said to have **ionic character**, e.g. H—Cl has a very polar covalent bond as a gas and becomes H$^+$ and Cl$^-$ when dissolved in water.

Bonding in compounds is always on a continuum from almost perfectly covalent to almost perfectly ionic.

Note: The electronegativities of C and H are very similar so the bond is a non-polar covalent bond.

Structure of Covalent Substances

Most covalent substances form simple molecules, e.g. I$_2$, C$_8$H$_{18}$, C$_{60}$

Simple Molecular Structures

Low Melting Point

Simple covalent molecules are held together by weak intermolecular forces, which take little energy to break and so have low melting points. They are gases, liquids

DAY 2

or low melting point solids. Where they are gases or liquids, the energy available at room temperature is sufficient to break the intermolecular bonds, e.g. iodine

Strong covalent bonds hold the iodine **atoms** together. Weak intermolecular forces (van der Waals) hold the iodine **molecules** together.

Electrical Non-Conductors

Simple covalent molecules have no free moving charged particles (ions or electrons) that can carry a current. All electrons are located within their atoms or as bonded pairs.

Soluble in Organic Solvents

Like dissolves like

Molecules without a dipole tend to be soluble in non-polar solvents. Most simple covalent molecules dissolve in organic solvents such as hexane. The strength of the bonding between solvent molecules is similar to the strength of the bonding between simple covalent molecules and the solvent.

Poor Solubility in Water

The strength of the bonds that can form between simple covalent molecules and water are not sufficient to overcome the hydrogen bonding between water molecules.

Polar covalent molecules, or those that can hydrogen bond, may be soluble in water, e.g. ethanol, glucose.

Giant Covalent Structures

Some covalently bonded substances form giant structures containing millions of atoms. These are known as macromolecules, giant lattices or networks, e.g. elements such as silicon or carbon in the form of graphite or diamond, and compounds such as silicon dioxide and silicon carbide.

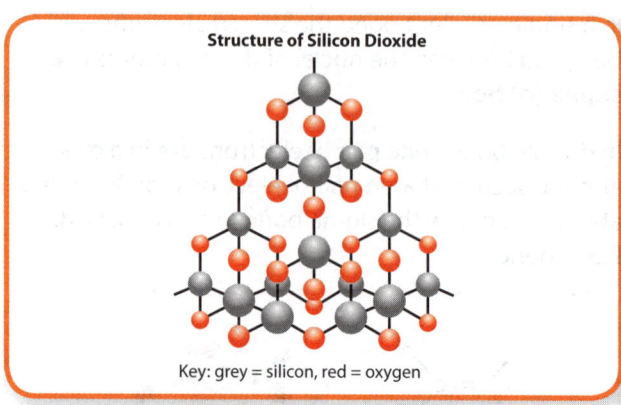

Structure of Silicon Dioxide
Key: grey = silicon, red = oxygen

There are no individual molecules; more atoms can always be added to the outside of the structure.

High Melting Point

Atoms in the lattice are held by strong covalent bonds, which require a large amount of energy to break.

Electrical Non-conductors

There are no charged particles that can move to carry a current. All the electrons are held tightly between the atoms.

Insoluble in Water and Organic Solvents

The bonds that would form between a solvent and the atoms of the lattice are not strong enough to overcome the strong covalent bonds in the lattice.

QUICK TEST

1. Define a covalent bond.
2. Draw a dot and cross diagram to show the bonding in ethene.
3. Give an example of a molecule that shows dative covalent bonding.
4. Explain why bromine is a liquid at room temperature.
5. Predict the polarity of a C—Br bond.
6. Explain the difference between a simple molecular structure and a giant covalent structure.

PRACTICE QUESTIONS

1. Which of the following is true of simple molecular structures? [1 mark]

   A They have high melting points.
   B They never contain metal atoms.
   C They form giant lattices.
   D They do not conduct electricity.

2. Which type of bonding is found in ethanoic acid? [1 mark]

   A Both σ bonds and π bonds
   B σ bonds only
   C π bonds only
   D Neither σ bonds nor π bonds

3. The exhaust fumes from a petrol engine contain a mixture of gases that may include water vapour, carbon dioxide, carbon monoxide and nitrogen oxides.

   a) Draw a dot and cross diagram for carbon dioxide. [2 marks]

   b) Carbon dioxide contains two different types of covalent bond. Name these bonds and explain the difference between them. [4 marks]

   c) Carbon monoxide has one dative covalent bond. What is meant by a dative covalent bond? [1 mark]

   d) Draw the structure of N_2O showing the dative covalent bond. [2 marks]

4. a) Explain why some covalent bonds have ionic character. [1 mark]

   b) Place the following bonds in order of increasing polarity: F—H, Br—H, Cl—Cl [1 mark]

5. Bromine is a liquid at room temperature but if some bromine is poured into a gas jar a brown vapour is seen to fill the jar.

   a) Use a dot and cross diagram to show why bromine forms diatomic molecules. [3 marks]

   b) Explain why bromine readily forms a vapour. [3 marks]

   Bromine has an electronegativity of 2.8 and forms a compound with fluorine that has an electronegativity of 4.0.

   c) Draw the structure of the compound that forms between bromine and fluorine. Label any polarisation. [1 mark]

   Hexane and water are two immiscible liquids that form two separate layers when mixed together. If an aqueous solution of bromine is mixed with hexane, most of the brown colour of the bromine is seen to transfer into the hexane.

   d) Explain why the bromine dissolves into the hexane rather than remaining in the water. [2 marks]

DAY 2 / 60 Minutes

Metallic Bonding and Structure/ Titration Techniques

Metallic Bonding and Structure

Metallic bonding is the strong electrostatic attraction between positive metal ions in a fixed lattice and delocalised electrons.

The metal atoms release their outer shell electrons to form a delocalised sea of electrons and a regular lattice of metal ions. The ions are held together by their mutual attraction to the delocalised electrons.

Strength of Metallic Bonding

The more outer shell electrons that are released to the delocalised sea of electrons, the higher the charge on the ion, and the stronger the attraction between the ions and electrons.

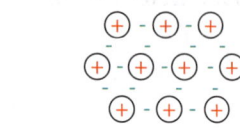
Sodium   Aluminium

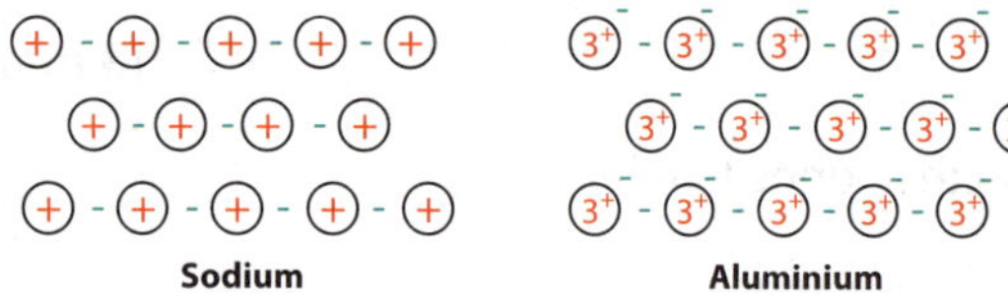

The smaller the ionic radius, the greater the attraction to the electrons and the stronger the metallic bond.

The stronger the metallic bond, the higher the melting point of the metal.

Increase in melting point of metals across a period: The strength of metallic bonding increases across a period because the charge on the ion increases and the radius of the ion decreases.

Decrease in melting point of metals down a group: The strength of metallic bonding decreases down a group because the ionic radius increases.

Note: The arrangement of the metal ions in the metallic lattice also influences the melting point so predictions about melting point from ionic radius and ionic charge alone are not always correct.

Properties of Giant Metallic Structures

All metals have giant structures.

High Melting Point

The melting point of most metals is high because of the strong attraction between delocalised electrons and metal ions. There are some exceptions, e.g. mercury and Group 1 metals.

Good Electrical Conductors

Metals conduct electricity as solids or liquids since the delocalised sea of electrons is able to move and carry the current when a potential difference is put across it.

Insoluble

Metals are insoluble in all solvents but will dissolve in mercury to form amalgams or in other metals to form alloys. Alloys are mixtures of metal atoms.

Malleable

Metals can be beaten into different shapes as solids because the layers of metal ions in the lattice can slide over each other in the delocalised sea of electrons. The metal atoms can settle into a new position without breaking the metallic bonding.

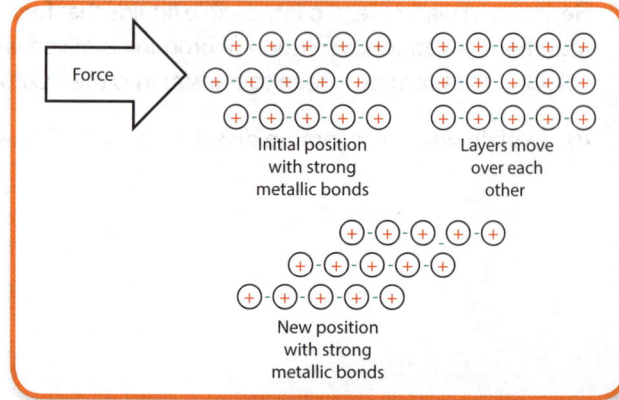

28

Titration Techniques

Titration is an analytical technique that can be used to find the quantity of substances in solutions. The substance to be determined (the analyte) reacts with a second solution (the titrant) for which both the concentration and volume are known. The reaction equation must be known.

Measured quantities of the titrant are slowly added to an accurately measured volume of the analyte until the reactants are in exactly the same ratio as is shown in the reaction equation. This is the end point of the titration. An indicator is used to show the end point.

The indicator will react with the first drop of titrant added after all the analyte has been up. It changes colour once it has reacted.

The exact volume of titrant added is measured and, since the concentration and volume of one reactant is known, the number of moles of the other reactant can be calculated.

Making Up a Volumetric Solution

One method of determining the quantity of a solid, e.g. the percentage of a substance in an impure sample, is to dissolve it in water and titrate. This requires an **accurately known mass** of the sample to be dissolved in a **precise volume** of solution.

1. Put a weighing boat on the balance and add approximately the quantity of analyte required. Record the mass as accurately as possible. The exact amount of the sample is not important but knowing exactly how much is being used is essential.

2. Empty the sample into a beaker and reweigh the weighing boat.

   Calculate the accurate mass of the sample being used by subtracting the final mass from the starting mass.

   This is known as **weighing by difference**.

3. Dissolve the sample in a moderate quantity of solvent (usually water, but sometimes dilute acid or an organic solvent). Use a glass rod to stir until the solution is completely transparent.

4. Use a glass funnel to pour the solution into a volumetric flask. Rinse the beaker with the solvent and pour the rinsings into the flask. Repeat twice. Rinse the glass rod into the flask. Rinse the glass funnel into the flask. Take care not to exceed the volume of the flask.

5. Make the volume in the flask up to the mark with solvent. Read the mark from the bottom of the meniscus with your eyes level with the mark. This avoids parallax errors caused by looking up or down at the solution. Stopper the flask and mix the contents by inverting the flask several times.

Performing a Titration

1. **Prepare the burette:** The burette is rinsed with the titrant to ensure that no water or other solutions will change the concentration of the titrant. Close the tap on the burette and pour in a small amount of the solution to be used (the titrant). Rinse the burette with the solution by removing it from the stand and inverting it carefully while covering the end. Discard the rinsings.

2. Move the burette to a position where it can be filled without having to pour above your head. Place an empty beaker under the jet of the burette. Check that the tap is closed. Place a burette funnel in the top of the burette. Pour some solution from the stock bottle of the titrant into a clean, dry beaker. Fill the burette from the beaker.

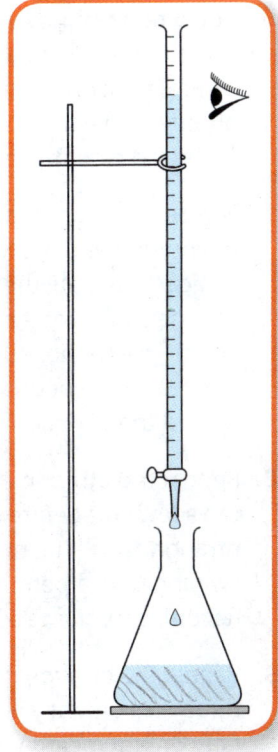

3. Place the filled burette on the bench. With the beaker still underneath, open the tap of the burette and allow solution to run through until there are no air bubbles in the jet or the tap.

29

DAY 2

4. Open the tap and run the solution down to the nearest whole number of cm^3 (for convenience) and record the reading. For a 50 cm^3 burette record to 0.05 cm^3. The smallest divisions are 0.1 cm^3 and the final figure is judged as either closest to the division mark or closest to halfway between the divisions.

5. **Prepare the conical flask:** Pour some of the test solution (analyte) into a clean, dry beaker. The sample is not pipetted directly from the whole sample to avoid any risk of contaminating the whole sample. Rinse a volumetric pipette with the solution. Transfer an exact amount of the test solution to a conical flask. When the solution has run out of the pipette, but still has a drop of solution in the tip, touch the tip against the side of the flask to remove the last drop.

6. Add **a few drops** of indicator to the conical flask. Excess indicator may change the titre. Place the conical flask on a white tile under the jet of the burette. The white tile allows the colour change of the indicator to be seen more easily.

7. Perform a rough titration by opening the tap of the burette and allowing solution to run into the conical flask until the indicator shows a colour change as the solutions first meet. Swirl the flask and then add more solution slowly from the burette until a permanent colour change is seen. Record the volume added.

8. Repeat the titration. This time close the burette tap a few cm^3 before the endpoint of the rough titration. Continue to add titrant **dropwise**, with swirling until a permanent colour change is seen. Record the volume added.

9. Repeat the titration until at least two and preferably three concordant values of volume are seen. Concordant means within 0.1 cm^3 of each other.

- The conical flask should be thoroughly rinsed between each titration but does not need to be dried.
- Either the titrant or the analyte can be in the burette.
- The burette can be read more easily if a white piece of paper is held behind it.

Indicators for Acid–Base Titrations

Different indicators are used for different titrations.

Titration	Indicator	Colour change
strong acid with strong base	phenolphthalein	colourless to pink
strong acid with weak base	methyl orange	orange to yellow
weak acid with a strong base	phenolphthalein	colourless to pink
weak acid with a weak base	no good indicator, so use a pH probe or data logger	

QUICK TEST

1. Draw a diagram of the structure of Mg.
2. What holds the atoms together in a metallic bond?
3. Why is the melting point of Na higher than that of K?
4. Why do metals conduct electricity as solids?
5. How many decimal places should be used to record the volume in a titration?
6. Which indicator is suitable for a titration of sulfuric acid with sodium carbonate?

PRACTICE QUESTIONS

1. The melting point of sodium is 336 K. Which of the following is true about the melting point of calcium? **[1 mark]**
   - A It will be less than 336 K because it releases more electrons to the delocalised sea of electrons. ☐
   - B It will be more than 336 K because it has a lower charge density. ☐
   - C It will be less than 336 K because it has a larger ionic radius. ☐
   - D It will be more than 336 K because it has fewer protons. ☐

2. When a flask containing a solution of hydrochloric acid is titrated against sodium hydroxide solution, which of the following is true? **[1 mark]**

	A suitable indicator is	With a colour change of
A	universal indicator	red to green
B	methyl orange	yellow to orange
C	phenolphthalein	pink to colourless
D	phenolphthalein	colourless to pink

3. Aluminium is a low density metal used in overhead cables that carry electricity.
   - a) Describe the structure and bonding in aluminium. **[2 marks]**
   - b) Explain why the structure of aluminium makes it a good conductor of electricity. **[2 marks]**

   Aluminium is also used extensively in the automobile industry because its low density gives good fuel consumption. Automobile companies have researched the possibility of replacing aluminium with magnesium.
   - c) Use your knowledge of the **structure and bonding** of metals to comment on the differences in properties you would expect between aluminium and magnesium. **[2 marks]**

4. An accurate titration result depends on following a careful procedure. Give the purpose of the following instructions for a titration.
   - a) Rinse the burette with the solution to be used. **[1 mark]**
   - b) Place a white tile under the conical flask. **[1 mark]**
   - c) Read the burette at eye level. **[1 mark]**
   - d) Swirl the flask while adding the titrant. **[1 mark]**
   - e) Touch the tip of the pipette against the side of the flask once it has drained. **[1 mark]**
   - f) Add the solution dropwise to the flask. **[1 mark]**
   - g) Continue titrating until three concordant values are obtained. **[1 mark]**

5. A student decided to measure the percentage by mass of citric acid in a sample of popping candy by performing a titration.
   - a) Describe how the student could make up a solution of popping candy that would be suitable to use in the titration. **[4 marks]**

   Citric acid is a weak acid.
   - b) Suggest a suitable substance to put in the burette for this titration and name a suitable indicator. **[2 mark]**
   - c) Comment on the suitability of this technique for measuring the citric acid content. **[1 mark]**

31

DAY 2 | 60 Minutes

Structures of Carbon/Shapes of Molecules

Structures of Carbon

Carbon most commonly exists in four different structures or allotropes.

Diamond

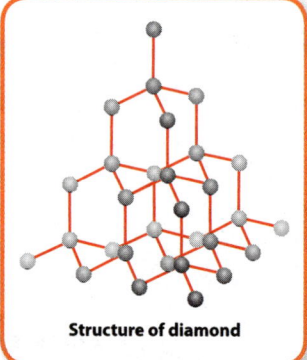
Structure of diamond

Bonding: covalent. Structure: giant (macromolecule).

Each carbon atom is covalently bonded to **four** other carbon atoms in a giant network.

Very hard with very high melting point because to move one atom requires the breaking of four strong bonds.

Non-conductor of electricity. There are no charged particles (ions or electrons) that are free to move in the lattice.

Graphite ‹ pencils

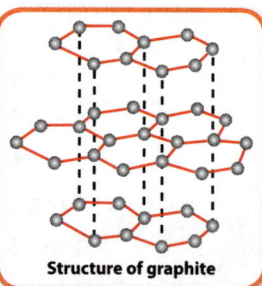
Structure of graphite

Bonding: covalent between atoms in the same layer; weak intermolecular forces between the layers (van der Waals). Structure: giant.

Each carbon atom is covalently bonded to **three** other carbon atoms to form a layer of hexagons. The fourth electron from each carbon atom is delocalised between the layers.

High melting point because for atoms to move requires three strong covalent bonds to be broken.

Conducts electricity because the delocalised electrons between the layers are able to move parallel to the layers when a potential difference is applied. A poor electrical conductor in comparison to most metals because electrons can only move in one plane.

Soft because the bonding holding the layers (planes of atoms) together is weak so the layers easily slide over each other.

Insoluble because the strong covalent bonds cannot be broken by the forces of attraction from solvent molecules.

Graphene

Bonding: covalent. Structure: giant but only one atom thick.

A two-dimensional giant structure that is a single layer of graphite hexagons. Each carbon has three σ bonds to three other carbons and one π bond. Graphene is a giant aromatic substance with an extended delocalised electron system.

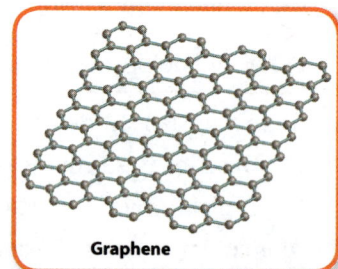
Graphene

Very strong. It has incredibly high tensile strength but is also brittle.

Conducts electricity due to the delocalised electrons, which are free to move.

Fullerenes

Bonding: covalent. Structure: simple. football

Each carbon atom is bonded to three others to form a mixture of hexagons and pentagons with double bonds. The structure is curved into enclosed spheres or open tubes.

E.g. Buckminsterfullerene C$_{60}$
The first fullerene to be discovered.
Soluble in organic solvents.

Found in soot and detected in space.

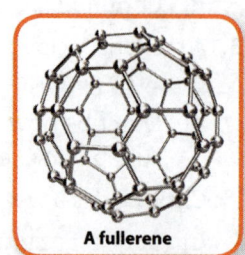
A fullerene

Non-conductor of electricity as electrons cannot easily pass from one molecule to another.

32

allotrope - different physical form of the same element
giant - large # of atoms covalently bonded.

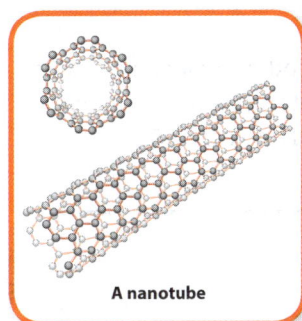
A nanotube

A **nanotube** is a **graphene** layer folded to produce a tube. They can be of varying diameter.

Electrical conductor as the delocalised electrons extend over the whole molecule and are free to move.

Fullerenes have proposed uses in a wide variety of fields such as: biomedicine for delivery of drugs; electronics due to their electrical conductivity; material science due to their strength, high thermal conductivity and ability to absorb infrared radiation; solar cells due to their ability to absorb ultraviolet radiation.

out of order

Shapes of Molecules

Electron Pair Repulsion Theory (EPRT)

(Also known as VEPRT: Valence Electron Pair Repulsion Theory.) Allows a prediction of the angle between two bonds and so the shape of molecules.

Regions of electron density around a central atom repel each other and will move as far apart as possible.

A region of electron density may be a single, double or triple bond or a lone pair of electrons.

Bond angles for single bonds around a central atom:

No. of bonds	Bond angle	Shape
2 bonds	180°	linear
3 bonds	120°	trigonal planar
4 bonds	109.5°	tetrahedral
5 bonds	90° and 120°	trigonal bipyramidal
6 bonds	90°	octagonal

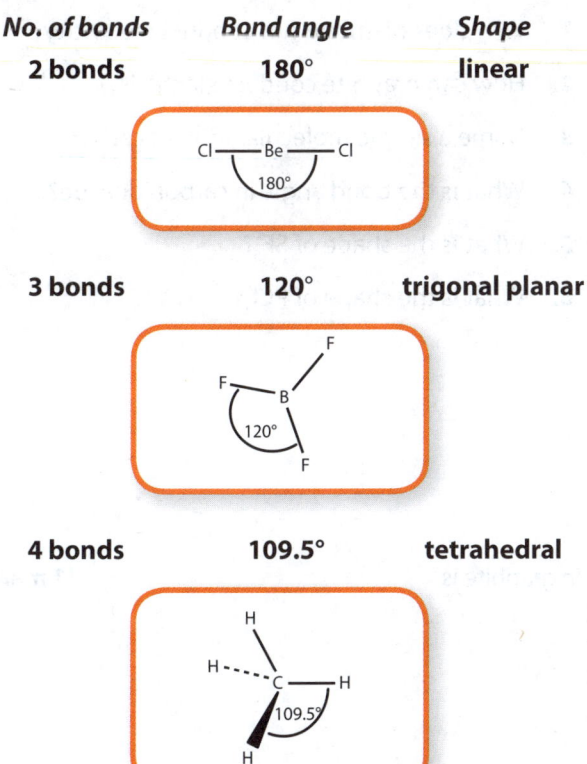

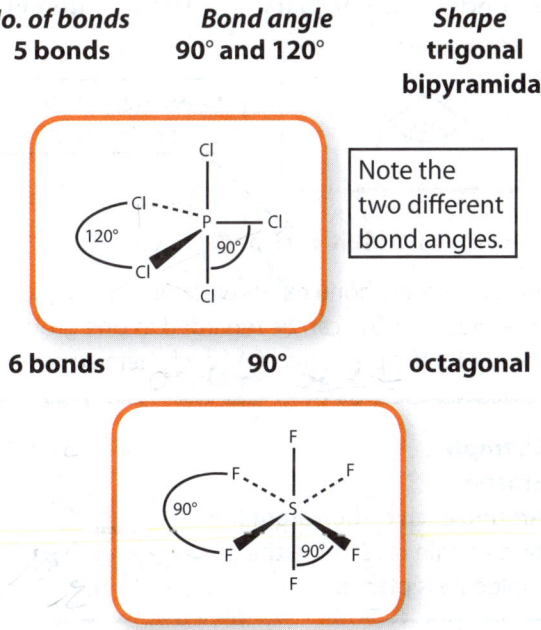

Note the two different bond angles.

Example
What is the shape of $SiCl_4$?

- Count the number of regions of electron density around the central atom = 4
- Decide the bond angle and shape using EPRT rules.
- Bond angle = 109.5°   So shape is tetrahedral.

Molecules with Lone Pairs of Electrons

Lone pairs of electrons around the central atom also repel other electron groups. The negative charge of lone pairs is concentrated closer to the central atom since they are not shared with another atom. This means they have a larger repulsion and reduce the bond angles between the bonding pairs. Each lone pair reduces the bond angle by approximately 2.5°.

33

DAY 2

Although the lone pair influences the bond angle, the overall shape of the molecule is given by the bonds.

No. of bonds | No. of lone pairs | Bond angle | Shape

3 bonds | 1 lone pair | 107° | pyramidal

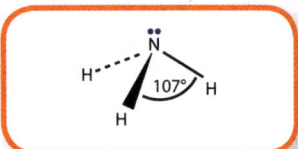

Note the bond angle is similar to methane but the shape is different because the lone pair do not count towards the shape.

2 bonds | 2 lone pairs | 104.5° | non-linear

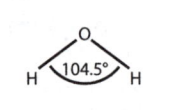

Sometimes called bent or V-shaped.

Molecules with Multiple Bonds

A double or triple bond repels electrons in a similar way to a single bond and can be regarded as one group of electrons for the purposes of A Level chemistry.

Example
Ethene
Around each carbon atom the bond angle is 120° and the molecule is planar.

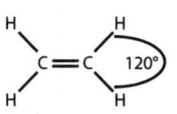

Example
Ethyne
Around each carbon atom the bond angle is 180° and the molecule is linear.

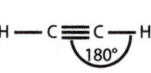

Molecular Ions

Ions follow the same rules as other molecules. Count the groups of electrons to determine the approximate bond angle. Reduce the bond angle by 2.5° for each lone pair.

Example
Carbonate ion
Two single bonds and one double bond, no lone pairs of electrons. Bond angle is 120° and the shape is trigonal planar.

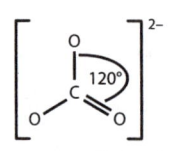

Example
Ammonium ion
Four single bonds, no lone pairs, bond angle 109.5° and shape tetrahedral.

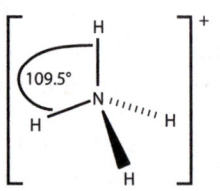

QUICK TEST

1. Why does diamond not conduct electricity?
2. How can graphite conduct electricity?
3. Name a simple molecular form of carbon.
4. What is the bond angle in carbon dioxide?
5. What is the shape of SF_6?
6. What is the shape of PCl_3?

PRACTICE QUESTIONS

1. The number of carbon–carbon bonds at each carbon atom in graphite is **[1 mark]**

   A 3   B 4

   C 2   D variable.

34

2. The bond angle of a molecule is determined by [1 mark]

   A   the number of covalent bonds only ☐

   B   electrons in bonds repelling each other ☐

   C   the number of lone pairs only ☐

   D   the number of regions of electron density. ☐

3. Buckminsterfullerene was discovered when a high energy laser was used to evaporate graphite in an inert atmosphere. Scientists discovered that it dissolved in an organic solvent to give a red solution and its formula was C_{60}.

   a) Why does it take a high energy laser to vaporise graphite? [2 marks]

   b) What evidence suggests that Buckminsterfullerene has a simple molecular structure? [2 marks]

   c) How does the bonding in Buckminsterfullerene differ from that in diamond? [2 marks]

   Later work resulted in the discovery of a wide variety of fullerenes including nanotubes, which are able to conduct electricity.

   d) Explain how a carbon nanotube is able to conduct electricity. [2 marks]

4. Ethylamine is an organic molecule used as a starting material for many industrial products. The structure of ethylamine is shown below.

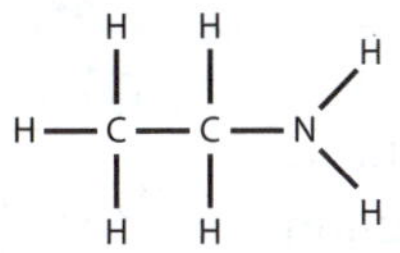

   a) Give the H—C—H bond angle in ethylamine. [1 mark]

   b) Explain how this bond angle can be estimated. [2 marks]

   c) What is the shape of the molecule around the carbon atoms? [1 mark]

   d) What is the H—N—H bond angle? [1 mark]

   e) Explain why the two angles are different. [2 marks]

5. Phosphorus can form two different chlorides, PCl_3 and PCl_5. The structure of PCl_5 is shown below.

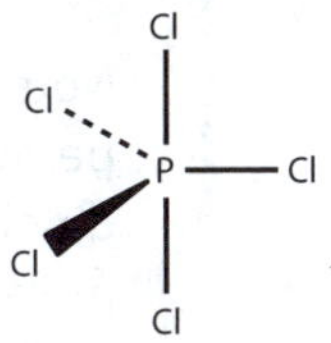

   a) Draw in the bond angles on the diagram and give a name to the shape of the molecule. [3 marks]

   b) PH_4^+ is known as a phosphonium ion. Suggest the shape of this ion and give a reason for your suggestion. [2 marks]

35

Intermolecular Forces

Intermolecular forces occur between molecules in simple structures. There are no intermolecular forces in giant ionic, giant covalent or giant metallic structures.

Intermolecular forces are caused by electrostatic attraction between partially charged particles and are less strong than chemical bonding.

The stronger the intermolecular forces, the higher the melting and boiling points of the substance.

Three main types of intermolecular forces:

- instantaneous dipole–induced dipole forces, also known as van der Waals forces, dispersion forces, and London forces
- permanent dipole–permanent dipole forces, also known as dipole–dipole forces
- hydrogen bonds.

Instantaneous Dipole–Induced Dipole Forces

These are found in all substances that have electrons. They are caused by a temporary, uneven distribution of electrons in a molecule that causes or induces a dipole in a neighbouring molecule.

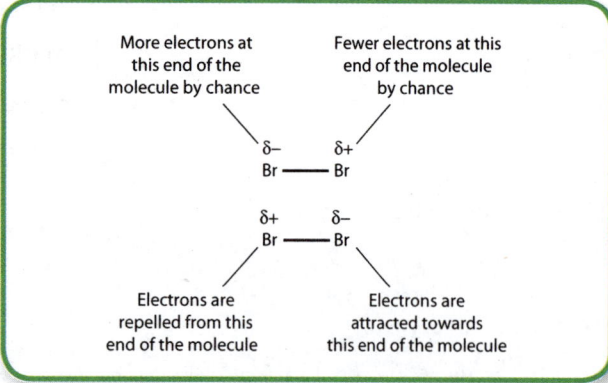

The strength of instantaneous dipole–induced dipole forces increases as the following factors increase:

- **number of electrons** in the molecule, as shown by the M_r, e.g.

$$I_2 > Br_2 > Cl_2 > F_2$$

- **chain length** of hydrocarbon chains

$$C_{10}H_{22} > C_6H_{14} > C_2H_6$$

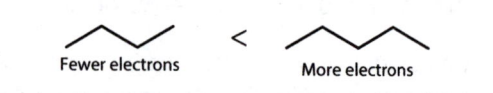
Fewer electrons < More electrons

- **area of contact** between molecules. More branched molecules have less area of contact. Molecules feel the electrostatic force more strongly if they can approach more closely, e.g.

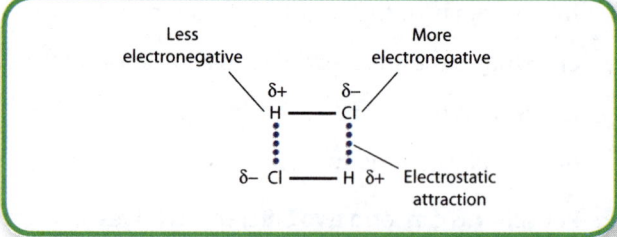
Same number of electrons, more spherical shape

Permanent Dipole–Permanent Dipole Forces

Dipole–dipole forces arise between molecules with a dipole caused by differences in electronegativity of the atoms.

Less electronegative   More electronegative
$\delta+$ $\delta-$
H —— Cl
$\delta-$ Cl —— H $\delta+$ Electrostatic attraction

Note: All molecules that show permanent dipole–permanent dipole forces **also** show instantaneous dipole–induced dipole forces. This means that a molecule with a dipole is likely to have a higher boiling point than one of the same M_r that does not have a dipole.

For the whole molecule to have a dipole there must be **both** polar covalent bonds and an asymmetric distribution of the charge. Polar molecules will all rotate to face the same direction in an electric field.

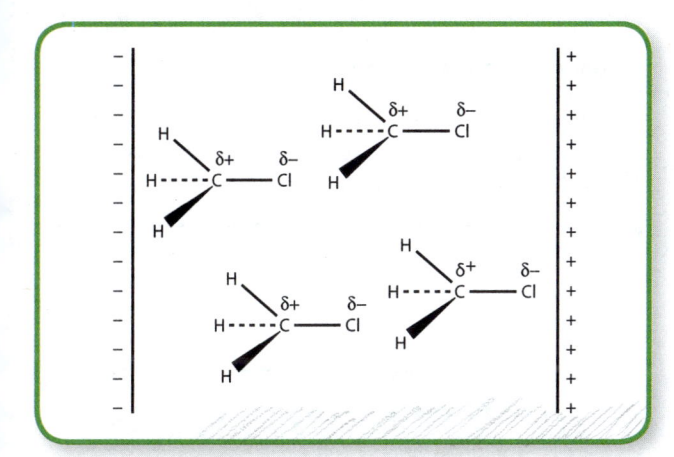

It is necessary to know the 3D shape of the molecule in order to decide if it has a dipole.

Example
Dichloroethene

Cis/Z isomer
Boiling point 60.3°C

Trans/E isomer
Boiling point 47.5°C

The cis isomer has an asymmetric distribution of polar covalent bonds and is therefore a polar molecule with both permanent dipole–permanent dipole forces and instantaneous dipole–induced dipole forces.

The trans isomer has a symmetrical distribution of polar covalent bonds and is therefore a non-polar molecule with only instantaneous dipole–induced dipole forces.

Hydrogen Bonding

Hydrogen bonding **only** arises where there is a H covalently bonded to F, N or O.

The bond between H and the very electronegative F, N or O causes the bond to become very polarised.

δ^-O—H$^{\delta+}$ The hydrogen nucleus shares very little of the electron so becomes relatively exposed. This very small hydrogen nucleus is then attracted to a lone pair of electrons on another atom. The small size of the H$^{\delta+}$ enables it to get very close to the lone pair forming a strong attraction known as a hydrogen bond.

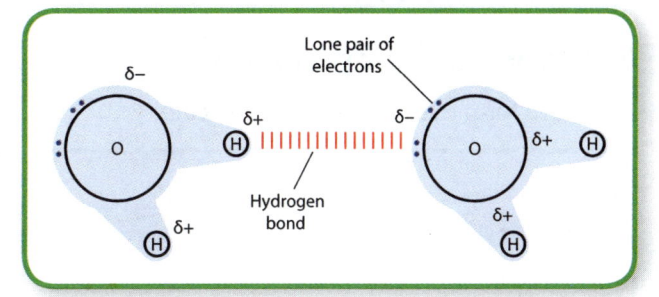

The hydrogen bond that forms is stronger than other intermolecular bonds, and is about as strong as $\frac{1}{10}$ of a covalent bond.

Molecules that can hydrogen bond have higher melting and boiling points than other molecules of the same M_r.

Organic groups that can hydrogen bond include alcohols, carboxylic acids, amines and amides.

Evidence for Hydrogen Bonding

The greater strength of hydrogen bonding over other intermolecular forces is demonstrated by the Group 6 hydrides.

From H_2S down to H_2Te the boiling points of Group 6 hydrides show the expected trend. Increasing numbers of electrons increases the instantaneous dipole–induced dipole forces giving a higher boiling point.

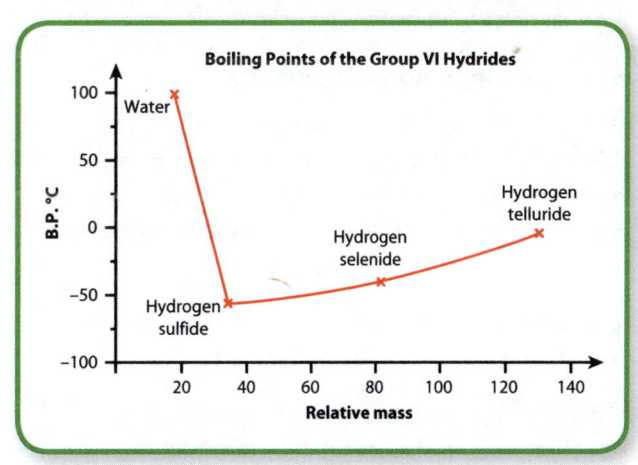

Boiling point $H_2Te > H_2Se > H_2S < H_2O$

H_2O breaks the trend and has the unusually high boiling point of 100°C because it has van der Waals force (instantaneous dipole–induced dipole forces) and hydrogen bonding between molecules.

37

DAY 3

Hydrogen Bonding in Water

Water can form two H bonds per molecule.

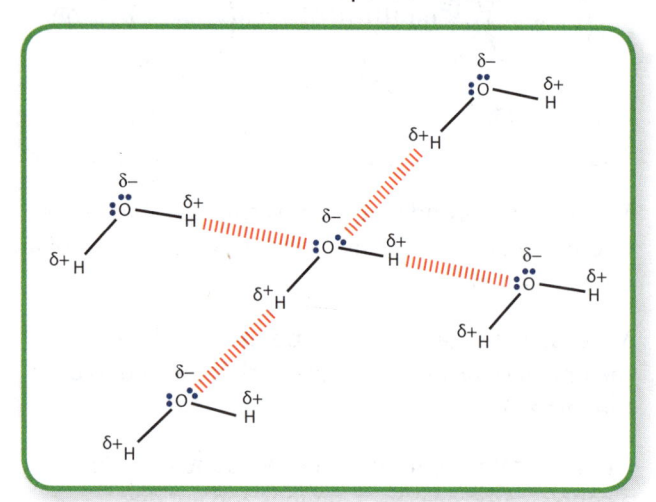

This gives water unusual properties: high boiling point, high specific heat capacity, high enthalpy of vaporisation and greater density as a liquid than as a solid.

In ice, each H_2O molecule forms two H bonds giving an open lattice structure with a low density. When ice melts some H bonds break allowing molecules to fall into the holes in the lattice, resulting in a smaller volume and higher density for liquid water. The highest density of water occurs at 4°C. *Cool!*

Hydrogen Bonding and Solubility

Molecules that can H bond are more likely to be soluble in water since they can hydrogen bond to the water molecules. The energy released when the H bonds form is sufficient to break the hydrogen bonding between water molecules.

Ions can form ion–dipole forces between the highly polar water molecules and a charged ion.

Molecules that cannot hydrogen bond and are not charged or polar are unlikely to be soluble in water as the strongest intermolecular bonds that can form between the water and the molecule are instantaneous dipole–induced dipole bonds. This does not release sufficient energy to disrupt the hydrogen bonding between water molecules.

Drawing Hydrogen Bonds

A diagram of a hydrogen bond always shows the partial charges on the atoms, the lone pair of electrons on the O, N, or F and the bond angle around the H as 180°.

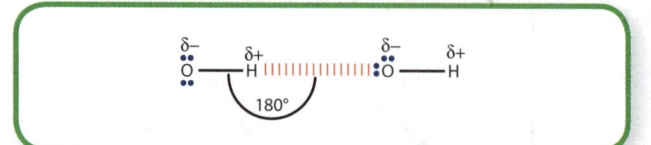

QUICK TEST

1. Why does the boiling point of elements in Group 7 increase down the group?

2. Which of the following would have the lowest boiling point and why?
pentane or 2,2-dimethylpropane

3. Which of the following are polar covalent bonds?

    C—F, C—H, C—N, C—O, I—F, S—H

4. Which of the following has polar covalent bonds but is not a polar molecule?

    CH_3F, H_2O, CCl_4, CO_2

5. Draw a diagram of hydrogen bonding in ammonia.

6. Why does propanone not form hydrogen bonds?

PRACTICE QUESTIONS

1. Which molecule has the strongest instantaneous dipole–induced dipole forces? **[1 mark]**

   A Methane ☐
   B Chloromethane ☐
   C Bromomethane ☐
   D Iodomethane ☐

2. Which of the following does not contain polar covalent bonds? **[1 mark]**

   A Chloroethane ☐
   B Propane ☐
   C Dichloropropene ☐
   D Propanone ☐

3. Which of the following can hydrogen bond? **[1 mark]**

   A Hydrogen fluoride ☐
   B Propanal ☐
   C Chloromethane ☐
   D Ethyl ethanoate ☐

4. Use ideas about intermolecular forces to explain why crude oil can be separated into fractions of different average molecular mass. **[3 marks]**

5. Ethanol can be oxidised to ethanal using acidified potassium dichromate. To obtain a high yield it is necessary to remove the ethanal from the reaction mixture as it forms. This prevents further oxidisation to ethanoic acid. The ethanal is removed as it forms by distillation so that the ethanal vaporises and is separated from the reaction mixture.

   ethanol   ethanal   ethanoic acid

   a) State the strongest type of intermolecular force that forms in pure ethanol, pure ethanal and pure ethanoic acid. **[3 marks]**

   b) Explain why ethanal will vaporise from the boiling reaction mixture but ethanol will not. **[2 marks]**

   c) In this reaction the ethanol and ethanoic acid dissolve well in the water-based potassium dichromate solution. Give a reason for this. **[2 marks]**

6. Explain why ice floats in water and draw a diagram to illustrate your answer. **[4 marks]**

39

DAY 3 — 60 Minutes

Energetics

Enthalpy Profile Diagrams

Enthalpy is heat energy and is measured in kJ mol^{-1}.

Exothermic Reactions Release Heat

The enthalpy of reactants is higher than the enthalpy of products. Energy is transferred to the surroundings during the reaction. The temperature of the surroundings increases.

Less energy is required to break bonds in reactants than is released when new bonds are formed in products. Heat is released; ΔH is negative.

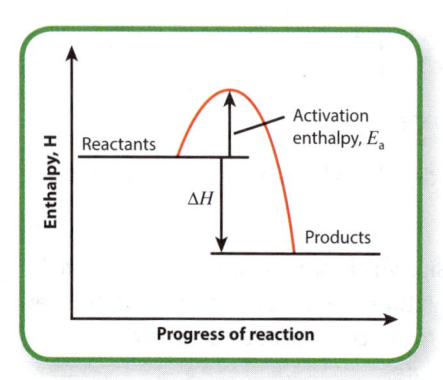

Endothermic Reactions Absorb Heat

The enthalpy of reactants is lower than the enthalpy of products. Energy is absorbed from the surroundings during the reaction. The temperature of the surroundings decreases. The activation energy is always high.

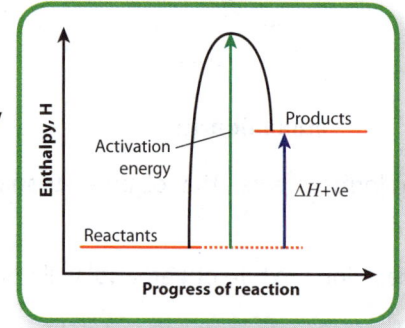

More energy is required to break bonds than is released when new bonds are formed; ΔH is positive.

Enthalpy level diagrams do not show activation energy.

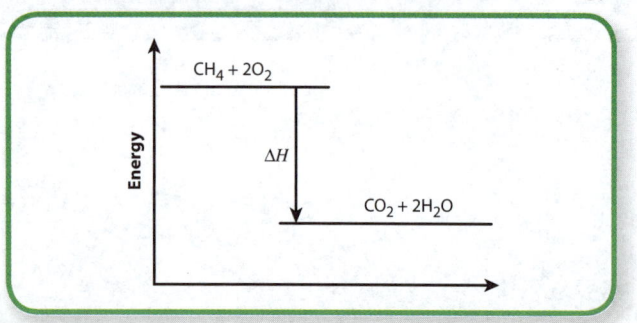

Standard Enthalpy Changes

Enthalpy change (ΔH) is the heat energy change measured at constant pressure.

Standard enthalpy changes ($\Delta H^\ominus$) are measured at 298 K and 1 atmosphere (100 kPa) with all reactants and products in their standard state.

Standard state is the phase (solid, liquid or gas) of the substance at 298 K and 1 atmosphere.

$\Delta H^\ominus$ **combustion** ($\Delta H^\ominus_c$) is the enthalpy change when one mole of a substance is completely burned in oxygen under standard conditions. This is always an exothermic reaction so enthalpies of combustion are always negative.

$\Delta H^\ominus$ **formation** ($\Delta H^\ominus_f$) is the enthalpy change when one mole of a substance is formed from its elements in their standard state under standard conditions.

$\Delta H^\ominus$ **neutralisation** ($\Delta H^\ominus_{neut}$) is the enthalpy change when one mole of water is formed in the neutralisation of an acid under standard conditions.

$\Delta H^\ominus$ **reaction** ($\Delta H^\ominus_r$) is the enthalpy change when the reaction given by the reaction equation occurs under standard conditions.

Bond enthalpy is the enthalpy change when one mole of specified bonds is broken from a molecule in its **gaseous** state.

Since enthalpy changes can be exothermic or endothermic they always require a sign.

Experiments to Measure $\Delta H^\ominus$

$\Delta H^\ominus$ Combustion

Burn a fuel in a spirit burner and capture the heat energy released in a calorimeter.

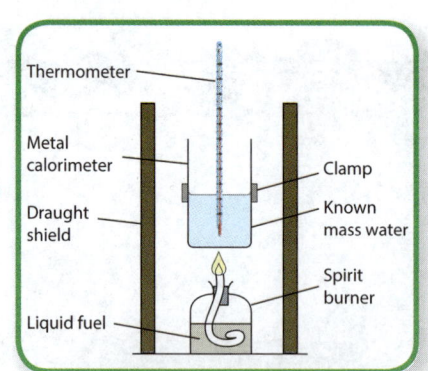

40

Measure
- starting temperature of water in calorimeter
- starting mass of fuel
- final temperature of water in the calorimeter
- final mass of fuel.

Calculate the energy change in the reaction:

$$Q = m \times c \times \Delta T$$

Energy change = Mass of water × specific heat capacity of water × change in temp of water

Convert this energy change in J to kJ by dividing by 1000 (standard enthalpy changes are quoted in kJ).

Q is the heat energy change in the surroundings of the reaction. In this case the surroundings are assumed to be the water in the calorimeter. Since there is a temperature **rise** this represents a **decrease** in enthalpy for the chemicals. The enthalpy change ΔH is negative.

Calculate the mass of fuel burned and then moles of fuel burned using moles = $\frac{mass}{M_r}$.

Divide the energy change by the number of moles of fuel burned to find the energy change in kJ mol^{-1}.

Example
A student carried out an experiment to measure the enthalpy change of combustion of propanol and obtained the following results:

	Starting	Final
mass of fuel + burner (g)	0.64	0.36
temperature of water (°C)	20.0	26.2
mass of water (g)	200	

Mass of fuel burned = 0.64 − 0.36 = 0.28 g

Temperature change = 26.2 − 20.0 = 6.2°C

Note: 1°C = 1 K

Specific heat capacity of water = 4.18 Jg^{-1}K^{-1}

$Q = m \times c \times \Delta T$

Energy change = 200 × 4.18 × 6.2 = 5183 J
= 5.183 kJ

M_r propanol $C_3H_8O = 60$

Moles of propanol burned = $\frac{0.28}{60}$ = 0.0047

Energy change per mole propanol = $\frac{5.183}{0.0047}$
= 1103 kJ

Enthalpy change of combustion = −1103 kJ mol^{-1}

The true enthalpy change of combustion of propanol is −2021 kJ mol^{-1}.

Why lab results may not match accepted results:
- Heat is lost to the surroundings rather than being transferred to the water.
- Heat is used to increase the temperature of the equipment.
- Incomplete combustion of fuel means the energy release is lower.
- Conditions are not standard.

Minimising experimental error:
- Use a cap on the spirit burner when not alight to reduce evaporative losses of fuel.
- Stir the water and measure the maximum temperature rise.
- Use a draft shield.

$\Delta H^\ominus$ Reaction for a Reaction in Solution

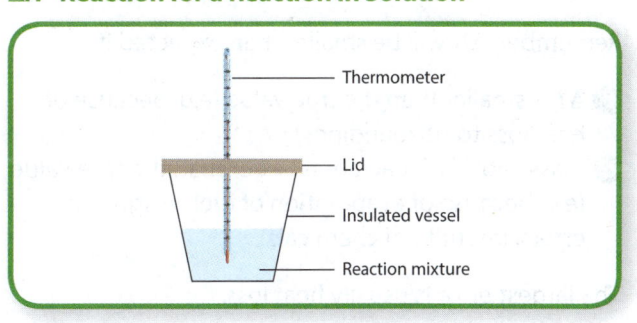

(Diagram: Thermometer, Lid, Insulated vessel, Reaction mixture)

Measure the temperature of reactant 1 every minute until there is no further change in temperature. At time = t add reactant 2 and measure temperature every minute from $t + 1$ onwards. Plot temperature against time and draw lines of best fit through the points before and after mixing.

DAY 3

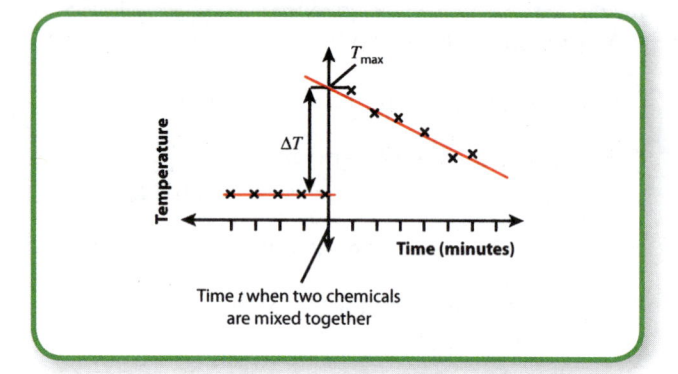

The temperature drop after mixing is due to heat loss to the surroundings. Extrapolating the line backwards allows an estimate of the temperature at time t taking into account heat loss to the surroundings. Measure the temperature difference before and after mixing (ΔT).

m is the mass of both reagents. Assume that all liquids have the same density as water (1 g cm^{-3}) unless told otherwise.

Assume that the specific heat capacity of the mixture is that same as that of water (4.18 J g^{-1} K^{-1}).

Use $Q = m \times c \times \Delta T$ to find energy change. Remember m is the mass of **both** reagents if two solutions are used.

Calculate the moles of the **limiting reagent** and divide the energy change by moles to find the energy change per mole.

Remember, ΔH will be smaller than expected if

- ΔT is smaller than the true value (e.g. because of heat loss to surroundings)
- mass/moles of reactant is larger than the true value (e.g. because of evaporation of fuel, weighing errors, impurity of chemicals).

The largest error is usually heat loss.

Don't forget

…in $Q = m \times c \times \Delta T$, m is the mass of water not the mass of the reactant.

…to give the sign for ΔH.

…to convert J to kJ for standard enthalpy changes.

QUICK TEST

1. Sketch an enthalpy profile diagram for the combustion of ethanol.

2. Which standard enthalpy change does the following reaction represent?

    $CH_4(g) + 2O_2(g) \longrightarrow CO_2(g) + 2H_2O(l)$

3. Why is this equation **not** $\Delta H^\ominus_f$ for butanol?

    $8C(s) + O_2(g) + 10H_2(g) \longrightarrow 2C_4H_9OH$

4. Write an equation for the standard enthalpy change of neutralisation of sulfuric acid with sodium hydroxide.

5. If the temperature rise of 20 g water in an enthalpy change experiment is 15°C, what is the energy change?

6. A student failed to put a lid on the container used in an enthalpy change of neutralisation measurement. What is the likely outcome on the value of $\Delta H^\ominus_{neut}$?

PRACTICE QUESTIONS

1. A reaction is described as endothermic if   [1 mark]

A	the reactants have a higher enthalpy than products	ΔH is negative	☐
B	the reactants have a higher enthalpy than products	ΔH is positive	☐
C	the reactants have a lower enthalpy than products	ΔH is negative	☐
D	the reactants have a lower enthalpy than products	ΔH is positive	☐

2. If burning 0.35 g of ethanol gives a temperature rise of 10°C in 250 g water, the standard enthalpy change of combustion of ethanol is   [1 mark]

A +2746 kJ mol^{-1} ☐   B −2746 kJ mol^{-1} ☐

C +1373 kJ mol^{-1} ☐   D −1373 kJ mol^{-1} ☐

3. A student uses a spirit burner to compare the enthalpy change of combustion of hexanol and hexane.

a) List the measurements that the student must take.   [3 marks]

b) The energy change was calculated from the experimental results with hexanol and found to be 500 kJ. If the mass of hexanol burned was 15 g, calculate the enthalpy change per mole.   [2 marks]

c) Suggest with reasons if this value is likely to be higher or lower than the true value.   [2 marks]

d) Suggest a modification to the experiment that could improve the accuracy of the results.   [1 mark]

e) The student repeated the experiment using hexane in the burner. During the experiment the outside of the calorimeter became very black. Suggest the reason for this blackening and suggest the effect on the final result.   [2 marks]

4. In an experiment to find the standard enthalpy of neutralisation for the reaction between 1 mol dm^{-3} potassium hydroxide and 1 mol dm^{-3} sulfuric acid the following results were obtained:

Volume of H$_2$SO$_4$ (cm^3)	Volume of KOH (cm^3)	Maximum temperature rise after mixing (°C)
25	25	6.8

Calculate the standard enthalpy of neutralisation.   [5 marks]

43

DAY 3 — 60 Minutes

Hess's Law

Hess's law states that the <u>enthalpy change of a reaction is independent of the route taken</u>, provided that reactants and products are in the same state.

Using $\Delta H^\ominus$ Combustion to Find the Enthalpy of a Reaction

Draw an enthalpy cycle.

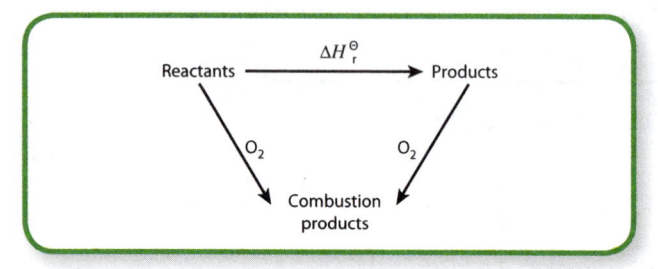

```
                ΔH°r
Reactants  ─────────────► Products
      ╲                  ╱
       ╲ O₂         O₂ ╱
        ╲             ╱
         ▼           ▼
          Combustion
           products
```

Hess's Law states that the energy change $\Delta H^\ominus_r$ directly from reactants to products is the same as the energy change going from **reactants** to combustion products and then from combustion products to **products**.

Energy change going from combustion products to **products** is the **negative** of $\Delta H^\ominus_c$ of the products.

Use data tables to find $\Delta H^\ominus_c$ for all reactants and all products.

Enthalpy change of reaction is the sum of all the $\Delta H^\ominus_c$ of reactants – sum of all $\Delta H^\ominus_c$ of products.

$$\Delta H^\ominus_r = +\Sigma\Delta H^\ominus_c \text{ (reactants)}$$
$$ -\Sigma\Delta H^\ominus_c \text{ (products)}$$

Example
Find the enthalpy of reaction for

$$C(s) + 2H_2(g) \longrightarrow CH_4(g)$$

Substance	$\Delta H^\ominus_c$ (kJ mol^{-1})
C(s)	−394
H$_2$(g)	−286
CH$_4$(g)	−890

Enthalpy cycle:

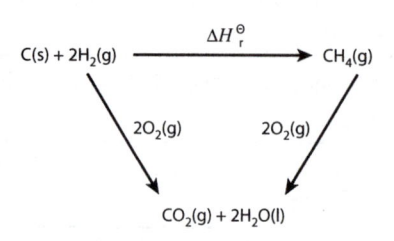

```
                  ΔH°r
C(s) + 2H₂(g) ─────────► CH₄(g)
        ╲                ╱
      2O₂(g)         2O₂(g)
          ╲            ╱
           ▼          ▼
          CO₂(g) + 2H₂O(l)
```

Sum of $\Delta H^\ominus_c$ reactants = $(-394) + 2(-286)$
$\phantom{\text{Sum of } \Delta H^\ominus_c \text{ reactants}} = -966$ kJ mol^{-1}

Sum of $\Delta H^\ominus_c$ products = -890 kJ mol^{-1}

$\Delta H^\ominus_r = (-966) - (-890) = -76$ kJ mol^{-1}

Note: Enthalpies of combustion are negative so the sum is always a negative value minus a negative.

Don't forget
...to multiply $\Delta H^\ominus_c$ by the number of moles in the reaction equation.

...to write state symbols on the enthalpy cycle.

Using $\Delta H^\ominus$ Formation to Find the Enthalpy of Reaction

Draw an enthalpy cycle.

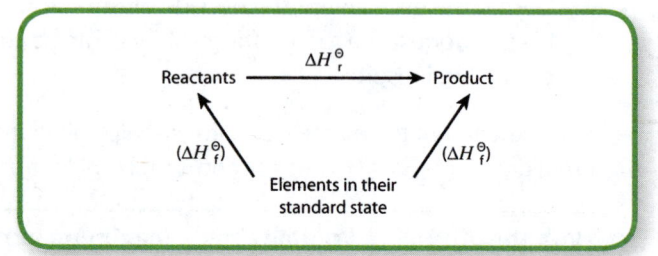

```
                ΔH°r
Reactants  ─────────────► Product
      ▲                  ▲
       ╲                ╱
       (ΔH°f)        (ΔH°f)
          ╲            ╱
         Elements in their
          standard state
```

Hess's Law states that the indirect route of reactants to elements and then elements to products has the same energy change as $\Delta H^\ominus_r$.

Enthalpy change going from reactants to elements is the negative of $\Delta H^\ominus_f$ of reactants.

44

Handwritten at top: $\Delta H_r = \Delta H_f \text{ products} - \Delta H_f \text{ reactants}$

Therefore the enthalpy of reaction is the negative of the sum of all enthalpies of formation of reactants + the sum of all enthalpies of formation of products.

$$\Delta H^\ominus_r = -\Sigma \Delta H^\ominus_f \text{ (reactants)}$$
$$+ \Sigma \Delta H^\ominus_f \text{ (products)}$$

Note: Enthalpies of formation are usually negative.

Example
Find the enthalpy of reaction for

$$CH_4(g) + 2O_2(g) \rightarrow CO_2(g) + 2H_2O(l)$$

Substance	$\Delta H^\ominus_f$ (kJ mol^{-1})
$CH_4(g)$	−75
$CO_2(g)$	−394
$H_2O(l)$	−286

Enthalpy cycle:

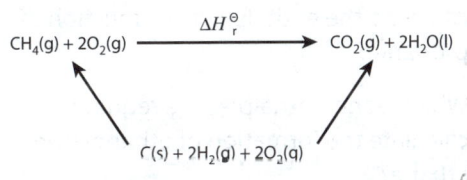

$CH_4(g) + 2O_2(g) \xrightarrow{\Delta H^\ominus_r} CO_2(g) + 2H_2O(l)$

with arrows from/to $C(s) + 2H_2(g) + 2O_2(g)$

Handwritten: what about O₂?

Sum of $\Delta H^\ominus_f$ reactants = −75 kJ mol^{-1}

Sum of $\Delta H^\ominus_f$ products = (−394) + 2(−286)
= −966 kJ mol^{-1}

$\Delta H^\ominus_r = -(-75) + (-966) = -891$ kJ mol^{-1}

Using Bond Enthalpies

Breaking bonds is endothermic; requires energy.
Forming bonds is exothermic; releases energy.

Bond enthalpy is the enthalpy change when 1 mole of bonds is **broken**. All bond enthalpies are positive values.

The enthalpy change when 1 mole of bonds is formed is the negative of the bond enthalpy.

The bond enthalpy for the same bond in a different molecule is slightly different. Mean bond enthalpy values are the average value for that bond across a range of molecules.

The enthalpy change of a reaction is the sum of the enthalpy change when bonds in the reactants are broken and the enthalpy change when bonds in the products are formed.

$$\Delta H^\ominus_r = \Sigma +\text{bond enthalpies (reactants)}$$
$$+ \Sigma -\text{bond enthalpies (products)}$$

Example
Find the enthalpy of reaction for

$$CH_4(g) + 2O_2(g) \rightarrow CO_2(g) + 2H_2O(g)$$

Bond	Mean bond enthalpy (kJ mol^{-1})
C—H	413
O=O	497
C=O	805
O—H	463

Sum of bond enthalpies of reactants
= 4(413) + 2(497) = +2646 kJ mol^{-1}

Sum of negative of bond enthalpies of products
= 2(−805) + 4(−463) = −3462 kJ mol^{-1}

Enthalpy of reaction = +2646 + (−3462)
= −816 kJ mol^{-1}

Don't forget
…to keep track of the double negative. It is the same as adding the values. Writing negative values for $\Delta H^\ominus_f$ in brackets may help.

…to write a sign and units on the answer.

Why calculations using bond enthalpies may not match other results:

- Bond enthalpies are averaged so may not be identical to a particular case.
- Bond enthalpies are calculated from the gaseous state; many reactions use other states, e.g. in the example water may be in the gaseous state but its standard state is liquid.

DAY 3

Using Experimental Results to Calculate Enthalpy of Reaction

Example
Calculate the enthalpy of hydration of calcium chloride.

$$CaCl_2(s) + 6H_2O(l) \rightarrow CaCl_2.6H_2O(s)$$

Measure the temperature change when $CaCl_2(s)$ and $CaCl_2.6H_2O(s)$ are dissolved and calculate the standard change of reaction for each.

$$\Delta H^\ominus_r = \frac{m \times c \times \Delta T}{\text{moles of chemical dissolved}}$$

Draw up an enthalpy cycle linking the reactions.

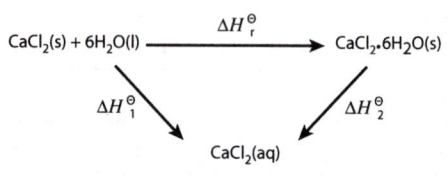

```
                    ΔH°_r
CaCl₂(s) + 6H₂O(l) ─────────→ CaCl₂.6H₂O(s)
         \                    /
          ΔH°₁              ΔH°₂
             \              /
              → CaCl₂(aq) ←
```

Use Hess's law to calculated the enthalpy of hydration $\Delta H^\ominus_1 - \Delta H^\ominus_2 = \Delta H^\ominus_r$

Percentage Uncertainty

Every measurement has some uncertainty. For example a mass that is given as 1.45 g could be any value between 1.4549 g and 1.4450 g because the balance only weighs to 2 decimal places.

The uncertainty or error in the measurement is +/− half of the smallest measurement that the measuring instrument can make. For equipment such as volumetric flasks, the error is calculated by the manufacturer and written on the side.

The percentage uncertainty or percentage error for any reading is given by $\frac{\text{error}}{\text{measurement}} \times 100\%$.

The percentage uncertainty for all measurements in an experiment can be added together to give a percentage uncertainty for the final value. The final value can then be quoted as +/− the percentage uncertainty. The number of significant figures quoted in the final value should reflect the percentage uncertainty.

QUICK TEST

1. Write the equation that represents the standard enthalpy of formation of water.
2. Why is the enthalpy of combustion of carbon the same as the enthalpy of formation of carbon dioxide?
3. Draw an enthalpy cycle to show how enthalpies of formation could be used to calculate the enthalpy of combustion of propanol.
4. Which bond enthalpies are required to calculate the formation of ethanol from ethane?
5. Why is the standard enthalpy of formation for O_2 zero?
6. What is the percentage uncertainty of a measurement of 20°C if the smallest division on the thermometer used is 0.5°C?

PRACTICE QUESTIONS

1. Which of the following represents $\Delta H^\ominus_f$ (ethanol)? [1 mark]

    A $C_2(s) + 6H(g) + O_2(g) \rightarrow C_2H_5OH(l)$ ☐

    B $2C(s) + 3H_2(g) + O(g) \rightarrow C_2H_5OH(g)$ ☐

    C $2C(s) + 3H_2(g) + 0.5O_2(g) \rightarrow C_2H_5OH(l)$ ☐

    D $4C(s) + 6H_2(s) + O_2(g) \rightarrow 2C_2H_5OH(l)$ ☐

2. Given the following bond enthalpies:

H—H	Cl—Cl	H—Cl
+436 kJ mol⁻¹	+242 kJ mol⁻¹	+432 kJ mol⁻¹

the standard enthalpy of formation of HCl is approximately [1 mark]

A +186 kJ mol⁻¹ ☐   **B** −186 kJ mol⁻¹ ☐   **C** +93 kJ mol⁻¹ ☐   **D** −93 kJ mol⁻¹ ☐

3. An enthalpy cycle for a reaction is shown below.

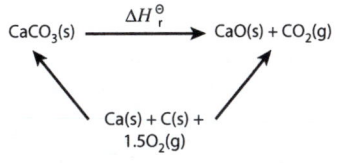

$$CaCO_3(s) \xrightarrow{\Delta H^\ominus_r} CaO(s) + CO_2(g)$$
$$Ca(s) + C(s) + 1.5O_2(g)$$

a) Apply Hess's Law to write an expression for $\Delta H^\ominus_r$ in terms of the other reactions. [1 mark]

b) Use the data below to calculate a value for $\Delta H^\ominus_r$. [3 marks]

Substance	$\Delta H^\ominus_f$ (kJ mol⁻¹)
CaCO₃(s)	−1207
CaO(s)	−635
CO₂(g)	−394

4. Standard enthalpy of combustion data can be used to calculate the standard enthalpy of formation of a compound.

a) Define the term standard enthalpy change of formation. [3 marks]

b) Use the data in the table to calculate the enthalpy of formation of propanone, CH₃COCH₃. Show all your working and draw an appropriate enthalpy cycle. [4 marks]

Substance	$\Delta H^\ominus_c$ (kJ mol⁻¹)
C (s)	−394
H₂(g)	−286
(CH₃)₂CO	−187

5.

Bond	Mean bond enthalpy (kJ mol⁻¹)
C—H	413
O=O	497
C=O	805
O—H	463
C—C	347

a) Use the bond enthalpies given to calculate the enthalpy of combustion for hexane. [3 marks]

b) The accepted value is −4163 kJ mol⁻¹. Suggest two reasons why this value differs from your answer in part **a)**. [2 marks]

DAY 3 — 60 Minutes

Kinetics

Collision Theory

Collision theory states that a reaction occurs when particles collide with sufficient energy.

Activation Energy

The activation energy E_a is the minimum energy with which particles must collide to produce a reaction. It is related to the energy required to break bonds in the reactants.

Higher Concentrations Increase Reaction Rate

Increasing the concentration of reactants increases the rate of reaction by increasing the frequency of collisions. There are more particles per unit volume so more collisions per second. This also applies to increasing the pressure of gases.

Higher Temperatures Increase Reaction Rate

Increasing the temperature increases the kinetic energy of the particles. This increases the percentage of the particles that have the activation energy for the reactions. It also increases the frequency of collisions.

Maxwell-Boltzmann Distribution of Energies at a Given Temperature

[Graph: Number of molecules with a given energy vs Energy. 60% of molecules in this range. E_a = activation energy of the reaction. The only molecules that can react are those with more energy than E_a.]

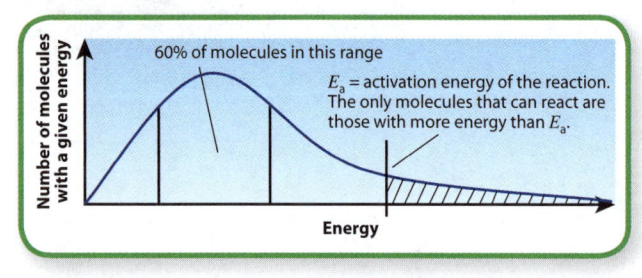

For a gas at a given temperature there is a range of energies that the gas molecules will have.

Most have an energy within a narrow range but some have higher or lower energies.

Only those collisions with energy ≥ (greater than or equal to) E_a will cause a reaction. For a typical reaction at room temperature this will be a small percentage of particles since E_a is typically on the right-hand side of the curve.

If carried out at the same temperature, reactions with a high E_a will have a slower rate of reaction than reactions with a lower E_a since a smaller percentage of collisions will be ≥ to E_a.

When the temperature increases, the average energy of the particles increases.

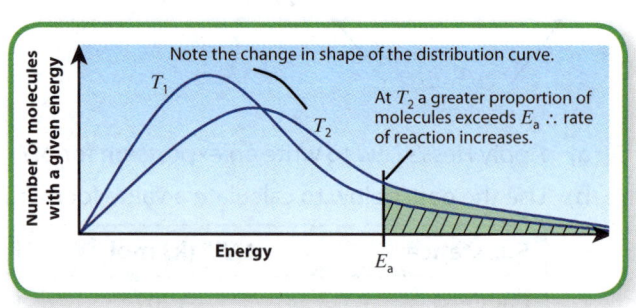

[Graph: Number of molecules with a given energy vs Energy, showing curves at T_1 and T_2. Note the change in shape of the distribution curve. At T_2 a greater proportion of molecules exceeds E_a ∴ rate of reaction increases.]

The area under the curve remains the same (this represents the number of particles) but the peak is lower and to the right.

Catalysts Increase the Rate of Reaction

A catalyst provides an alternative route through the reaction that has a lower E_a.

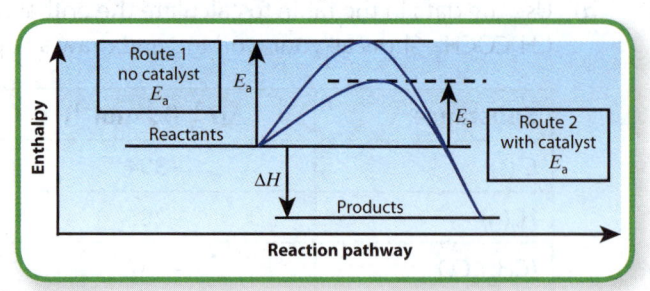

[Graph: Enthalpy vs Reaction pathway. Route 1 no catalyst E_a. Route 2 with catalyst E_a. Reactants, Products, ΔH.]

A lower E_a means that a higher percentage of the particles have ≥ E_a.

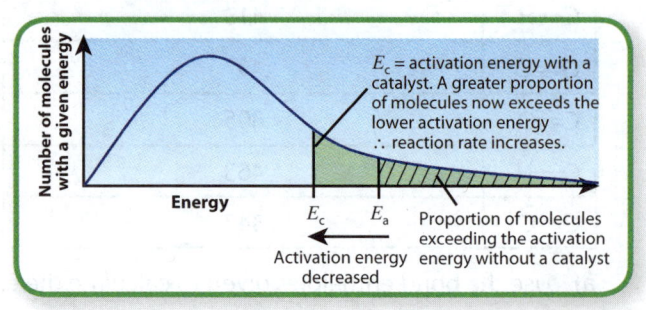

[Graph: Number of molecules with a given energy vs Energy. E_c = activation energy with a catalyst. A greater proportion of molecules now exceeds the lower activation energy ∴ reaction rate increases. E_c, E_a. Activation energy decreased. Proportion of molecules exceeding the activation energy without a catalyst.]

A catalyst is not used up by the reaction so a small amount can catalyse many reactions.

adhesion of molecules to a surface

A **heterogeneous catalyst** is in a different phase from the reactants, often a solid catalyst and gaseous reactants, e.g. a catalytic convertor in a car has a solid metal catalyst and gaseous reactants.

- Reactant particles are adsorbed onto a solid catalyst surface.
- The bonds between the reactant atoms weaken and break. The reactant particles are held in a favourable position to react.
- New bonds form between the product atoms.
- The products desorb and diffuse away from the catalyst surface allowing new reactants to take their place.

The larger the surface area, the more effective the catalyst.

Solid Catalysts Can Be Poisoned

The surface of a heterogeneous catalyst can become poisoned. This is when the surface is blocked by a substance that is adsorbed permanently to the surface. This prevents access to the reactants and so the catalyst stops working, e.g. lead additives in petrol poison the catalysts in catalytic convertors.

Homogeneous catalysts are in the same phase as the reactants, usually aqueous. Often an intermediate is formed that goes on to form the product and regenerate the catalyst.

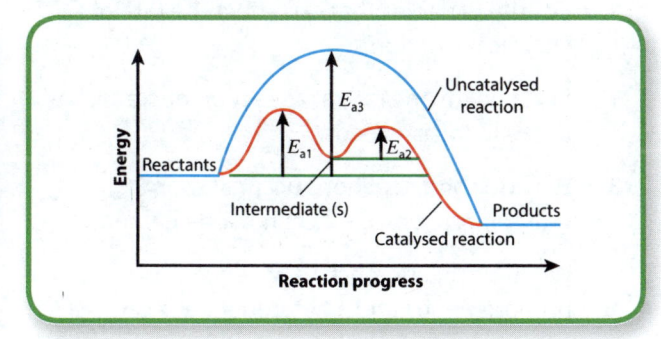

Following the Rate of a Reaction

The rate of a reaction can be followed by monitoring the appearance of products or the disappearance of reactants over time.

Measuring a Gaseous Product with Time

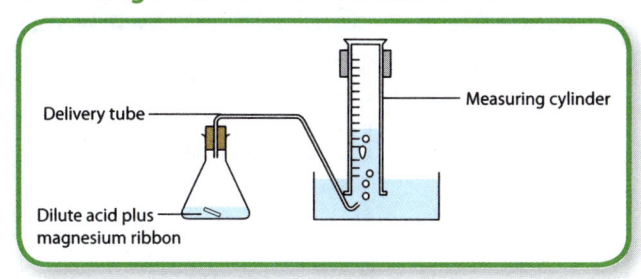

Problems: If the gas is water soluble then the true volume will be higher than the measured volume until the water is saturated with the gas.

The gas emerges as separate bubbles and so is not continuous. If a measurement is taken immediately before a bubble emerges then it will be lower than the true value.

There must be enough pressure in the flask to allow the bubbles to rise through the water. There is a delay before any volume is recorded.

A gas syringe can be used to overcome these problems.

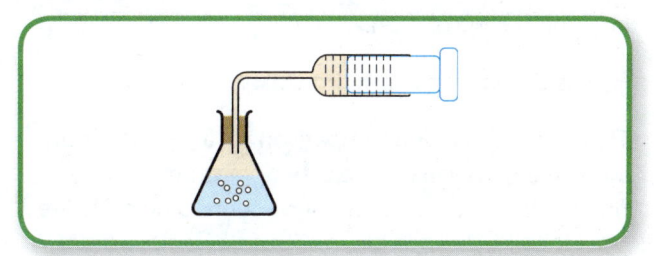

Measuring Mass Loss from a Gaseous Product

Measure the mass at timed intervals.

A cotton wool plug prevents liquids escaping.

Measuring Appearing or Disappearing Colour

If a reactant or product is coloured the colour can be measured at times intervals using a colorimeter.

Finding the Rate from the Time vs Amount Graph

Plot the data of volume/mass loss/change in absorbance against time.

The gradient at $t = 0$ gives the initial rate of reaction.

DAY 3

For reaction A ⟶ B.

Concentration of A vs Time graph

Concentration of B vs Time graph

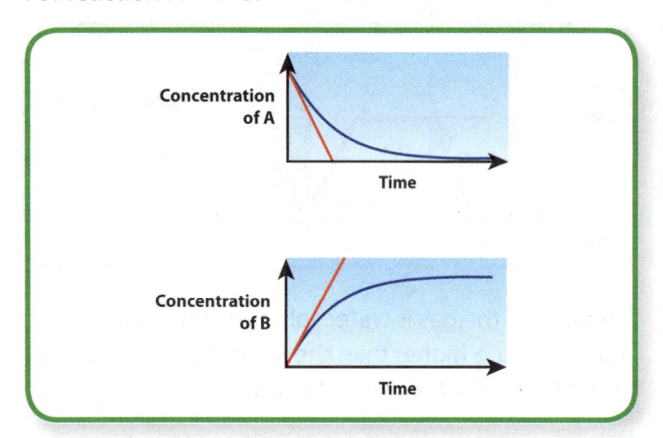

Measuring Appearance of a Solid Product

As the solid collects it obscures the cross as seen from above. Measure the time at which the cross disappears completely.

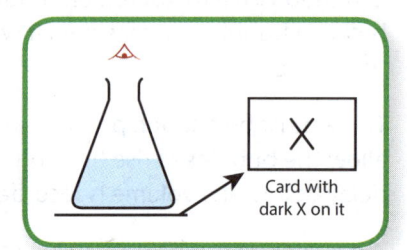

Diagram: flask above card with dark X on it

To compare results, always use the same flask, cross and volume of solution.

Rate is proportional to 1/time taken.

This method allows a comparison of rate for reactions with changing variables, e.g. temperature, but does not allow the absolute rate of reaction to be calculated.

Finding the Rate from a Clock Reaction

A clock reaction measures the time taken to get to a certain fixed point in the reaction, when a certain amount of product has been made. It allows rates of reaction to be compared taking only one measurement.

> **Example**
> Measuring the rate of oxidation of iodide ions by hydrogen peroxide (a slow reaction)
>
> $H_2O_2(aq) + 2I^-(aq) + 2H^+(aq) \longrightarrow I_2(aq) + 2H_2O(l)$
>
> When the reactants are mixed together a small known amount of thiosulfate is added as well. This reacts with any iodine produced in an instantaneous reaction.
>
> $2S_2O_3^{2-}(aq) + I_2(aq) \longrightarrow S_4O_6^{2-}(aq) + 2I^-(aq)$

No free iodine is seen until all the thiosulfate is used up. The iodine suddenly appears and the time is recorded. To make it easier to see when the iodine appears starch is added at the start and the reaction turns blue-black as soon as the free iodine appears.

The rate is proportional to 1/time taken.

The absolute rate of reaction can be calculated since the number of moles of iodine needed to react with the added thiosulfate can be calculated.

Controlling Variables

The independent variable is what is being changed, e.g. temperature, concentration, catalyst.

The dependent variable is what is being measured in the rate reaction, e.g. volume, mass, absorbance.

Since a number of factors affect the rate of reaction it is essential to control the value of all variables other than the independent variable, e.g. keep the concentrations the same when investigating temperature.

QUICK TEST

1. List the variables that can affect the rate of a chemical reaction.
2. Explain why the rate of a reaction decreases as the reaction proceeds.
3. Why do some reactions occur at room temperature while others must be heated?
4. What is the difference between a homogeneous and a heterogeneous catalyst?
5. Suggest a way to follow the rate of reaction for magnesium with hydrochloric acid.
6. Sketch a graph showing how the rate of reaction could be found from a time vs concentration graph.

PRACTICE QUESTIONS

1. The rate of the reaction between potassium iodide and hydrogen peroxide increases when the concentration of potassium iodide is increased. This is because [1 mark]

   A the kinetic energy of the particles increases

   B the collision frequency increases

   C the number of particles with E_a or greater increases

   D the value of E_a decreases.

2. Which of the following will **not** increase the rate of a gas reaction? [1 mark]

   A Increasing the temperature   B Adding a catalyst

   C Increasing the volume   D Increasing the pressure

3. The graph below shows the distribution of energies of particles of a gas at 298 K.

   a) Sketch a further line on the graph showing how the shape of the curve would change at 308 K. [3 marks]

   b) Use the two lines on the graph to explain why increasing the temperature increases the rate of a reaction. [2 marks]

   c) Increasing the concentration of reactants also increases the rate of a reaction. Suggest what effect increasing the concentration would have on the shape of the curve. [2 marks]

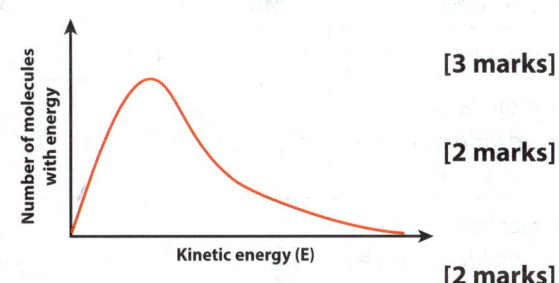

4. $CuSO_4$(aq) acts as a catalyst in the following exothermic reaction:

   $$2I^-(aq) + S_2O_8^{2-}(aq) \rightarrow I_2(aq) + 2SO_4^{2-}(aq)$$

   a) Explain what is meant by a catalyst. [2 marks]

   b) What **type** of catalyst is $CuSO_4$(aq)? [1 mark]

   c) Draw a labelled energy profile diagram showing the course of the reaction with and without the $CuSO_4$(aq). [4 marks]

   d) Describe a method that could be used to calculate the initial rate of this reaction. [4 marks]

5. Catalytic convertors are fitted to car exhaust pipes to reduce the quantity of polluting gases emitted. They are typically a thin layer of platinum/rhodium on a ceramic honeycomb structure. In the first few minutes after the engine is started the emissions are high but soon drop as the engine continues to run.

   a) Why are the metal catalysts put onto a honeycomb ceramic structure? [2 marks]

   b) Why are the polluting emissions high when the engine is first started? [1 mark]

   c) What **type** of catalyst is a catalytic convertor? [1 mark]

   d) Explain how a reaction proceeds when a catalyst like the one used in a catalytic convertor is used. [4 marks]

DAY 4 — 60 Minutes

Ar – Relative atomic mass

Calculations 2

Empirical Formula

The empirical formula is the simplest whole number ratio of elements in a compound. It can be calculated from a known mass of elements in the compound.

1. Divide the mass of each element by its A_r to find the moles of that element in the compound.
2. Divide the moles of each element by the smallest number of moles to find the ratio of moles.
3. Multiply the ratio of moles until all values are whole numbers.

Element	Mg	S	O
mass (g)	36.45	48.15	96.00
mole ratio – divide by atomic mass	$\frac{36.45}{24.3} = 1.5$	$\frac{48.15}{32.1} = 1.5$	$\frac{96.00}{16} = 6$
element ratio – divide largest by smallest	$\frac{1.5}{1.5} = 1$	$\frac{1.5}{1.5} = 1$	$\frac{6}{1.5} = 4$
empirical formula		$MgSO_4$	

Molecular Formula

The molecular formula is the number of atoms of each type of element in each molecule. It could be the same or a multiple of the empirical formula. For example, ethene C_2H_4 and hexene C_6H_{12} both have the empirical formula CH_2.

The molecular formula can be found from the empirical formula and the molecular mass.

1. Find the relative mass of the empirical formula by adding together all the atomic masses of the elements in the formula.
2. Find how many empirical formulae there are in the molecular formula by dividing the molecular mass by the mass of the empirical formula.
3. Find the molecular formula by multiplying the empirical formula by this factor.

Example
What is the molecular formula of a molecule with empirical formula CH_2 and M_r 84?

Mass of empirical formula: $12 + 2 = 14$

Divide molecular mass by mass of empirical formula: $\frac{84}{14} = 6$

Multiply the empirical formula by this factor:
$6 \times CH_2 = C_6H_{12}$

Percentage Yield

The percentage yield is the percentage of desired product made compared to the maximum that could be made. Reasons why the yield may not be 100% include side reactions that give a different product, equilibria systems, which always contain some reactants, or some of the product is lost when transferring from one piece of equipment to another.

$$\% \text{ yield} = \frac{\text{actual yield}}{\text{theoretical yield}} \times 100\%$$

1. Calculate the mass of the maximum possible yield from the mass of reactants and the reaction equation (see page 16).
2. Divide the actual mass of product by the maximum yield and multiply by 100%.

Example
What is the percentage yield of ammonium sulfate if the starting mass of ammonia is 100 g and the final mass of ammonium sulfate is 364 g?

$2NH_3(aq) + H_2SO_4(aq) \longrightarrow (NH_4)_2SO_4(aq)$

1. Calculate the maximum mass from the equation:
$M_r \, NH_3 = 17$
$\frac{100}{17} = 5.88$ starting moles of ammonia
Mole ratio $= 2 : 1$
Moles of $(NH_4)_2SO_4 = 0.5 \times 5.88 = 2.94$
M_r of $(NH_4)_2SO_4 = 132.1$
Maximum mass $(NH_4)_2SO_4 = 132.1 \times 2.94 = 388$ g

2. Divide the actual mass of product by the maximum yield and multiply by 100%:
$\frac{364}{388} \times 100\% = 93.7\%$

Atom Economy

The atom economy shows the % by mass of the reactant atoms that become useful products. It measures the amount of waste in a reaction.

Atom economy = $\frac{\text{sum of } M_r \text{ of desired products}}{\text{sum of } M_r \text{ of reactants}} \times 100\%$

Example
Manufacture of chloroethane from ethanol

$C_2H_6O + HCl \longrightarrow C_2H_5Cl + H_2O$

M_r of desired product $C_2H_5Cl = 64.5$

$M_r \ C_2H_6O = 46$ $M_r \ HCl = 36.5$

Atom economy = $\frac{64.5}{82.5} \times 100\% = 78.2\%$

Synthetic routes should be designed to maximise the incorporation of all materials used in the process into the final product.

Gas Volumes

All gases occupy the same volume at the same temperature and pressure.

This means that for an equation involving gases the mole ratio in the reaction equation is the same as the volume ratio for the gases, regardless of the temperature and pressure.

To find the volume of any gas in the reaction when the volume of another gas is known, multiply by the mole ratio.

Example
If 25 cm³ of nitrogen was mixed with oxygen, what volume of nitrogen oxide would be made?

$N_2(g) + O_2(g) \longrightarrow 2NO(g)$

Mole ratio $N_2 : NO = 1 : 2$

$25 \times 2 = 50 \ cm^3$ of NO

Gas Volumes to Moles

At room temperature and pressure (rtp) the volume of one mole of any gas is 24 dm³.

To convert from moles to volume multiply by 24.

To convert from volume gas at rtp to moles divide by 24.

Example

$CaCO_3(s) \longrightarrow CaO(s) + CO_2(g)$

If 250 kg of $CaCO_3$ was heated to produce CaO, what volume of $CO_2(g)$ would be released?

To convert from kg to g multiply by 10^3.

$M_r \ CaCO_3 = 100$ moles in 250 kg = $\frac{250}{100}$
$= 2.5 \times 10^3$

Mole ratio $CaCO_3 : CO_2 = 1 : 1$ moles $CO_2 = 2.5 \times 10^3$

To convert from moles to volume multiply by 24.

Volume $CO_2 = 2.5 \times 10^3 \times 24 = 60\ 000 \ dm^3$

Air at rtp is 21% oxygen by volume. How many moles of oxygen are present in a room with a volume of 72 000 dm³?

21% of 72 000 dm³ is $\frac{72\ 000}{100} \times 21 = 15\ 120 \ dm^3$

To convert from volume gas at rtp to moles divide by 24.

$\frac{15\ 120}{24} = 630$ moles oxygen

DAY 4

The Ideal Gas Equation

The ideal gas equation shows the relationship between the number of moles (n), volume (V), temperature (T) and pressure (p) of a gas.

$$pV = nRT$$

R is the universal gas constant and will be provided. For SI units it is 8.31 J K^{-1} mol^{-1}.

Correct Units Are Essential

Pressure, p, is in Pascals.
Pressures are often quoted in kPa. To convert into Pa multiply by 1000.

Volume, V, is in m^3.
Chemists usually use dm^3 or cm^3. To convert dm^3 into m^3 divide by 1000. To convert cm^3 into m^3 divide by 1 000 000.

Temperature T is in Kelvin.
Chemists often use °C. To convert °C to K add 273.

Using the Ideal Gas Equation

1. Rearrange the equation to make the required answer the subject of the equation.
2. Convert all information into the standard units.
3. Substitute values into the equation.

Example
How many moles of gas are present in 200 cm^3 of gas at 70°C and 101 kPa?

Rearrange the equation to make the required answer the subject of the equation:

$$pV = nRT \qquad n = \frac{pV}{RT}$$

Convert all information into the standard units:

200 cm^3 = $\frac{200}{1\,000\,000}$ = 2 × 10^{-4} m^3

70°C = 70 + 273 = 343 K

101 kPa = 101 × 1000 = 101 000 Pa

Substitute values into the equation:

$n = \frac{101\,000 \times 0.0002}{8.31 \times 343}$ = 0.0071 moles of gas

---

Finding the M$_r$ of a Gas

Combining the ideal gas equation with moles = $\frac{mass}{M_r}$

$$p \times V = \frac{mass \times R \times T}{M_r} \quad \text{then} \quad M_r = \frac{mass \times R \times T}{p \times V}$$

Example
What is the M$_r$ of a gas if 12 g occupies 12.4 dm^3 at 298 K and 150 kPa?

Convert all information into the standard units:

12.4 dm^3 = 0.0124 m^3

150 kPa = 150 000 Pa

Substitute values into the equation:

$M_r = \frac{mass \times R \times T}{p \times V} = \frac{12 \times 8.31 \times 298}{150\,000 \times 0.0124} = 16.0$

QUICK TEST

1. Calculate the empirical formula of a hydrocarbon containing 24 g carbon and 4 g hydrogen.

2. State the molecular formula of a compound with empirical formula C$_2$H$_4$O and M$_r$ 88.

3. Calculate the percentage yield if 12.0 g of Cu are made from 19.1 g of CuS.

   CuS(s) + O$_2$(g) $\longrightarrow$ Cu(s) + SO$_2$(g)

4. Calculate the atom economy for the production of ethene from ethanol.

   CH$_3$CH$_2$OH $\longrightarrow$ CH$_2$CH$_2$ + H$_2$O

5. Calculate the volume of oxygen produced from the decomposition of 0.2 moles of hydrogen peroxide at rtp.

   2H$_2$O$_2$(aq) $\longrightarrow$ 2H$_2$O(l) + O$_2$(g)

6. Calculate the M$_r$ of a gas that has a volume of 24 dm^3, a pressure of 101 kPa and a mass of 2 g at 293 K.

PRACTICE QUESTIONS

1. A compound contains 50% oxygen and 50% sulfur by mass. The empirical formula of the compound is [1 mark]

   A   SO ☐
   B S_2O ☐
   C SO_2 ☐
   D S_2O_2 ☐

2. A hydrocarbon has a M_r of 112 and an empirical formula of CH_2. Which of the following could **not** be the hydrocarbon? [1 mark]

   A   Cyclooctane ☐
   B   2-methylheptene ☐
   C   1,2-dimethylcyclohexane ☐
   D   3-methylcycloheptene ☐

3. A 10.0 g sample of a compound contains 4.0 g of carbon, 0.7 g of hydrogen and the remainder is oxygen. The empirical formula of the compound is [1 mark]

   A $C_2H_4O_2$ ☐
   B $C_4H_7O_5$ ☐
   C CH_2O ☐
   D C_2H_5O ☐

4. A student completely oxidised a piece of magnesium by burning it in air. The starting mass of the metal was 2.50 g and the mass after burning was 4.14 g.

   a) What mass of oxygen from the air was used? [1 mark]
   b) Deduce the formula of the oxide. [3 marks]

5. Ethyl ethanoate can be made via two different routes.

   Route 1: $CH_3CH_2OH + CH_3COOH \longrightarrow CH_3CH_2OCOCH_3 + H_2O$
   Route 2: $CH_3CH_2OH + CH_3COCl \longrightarrow CH_3CH_2OCOCH_3 + HCl$

   a) Which route has the higher atom economy? [1 mark]
   b) If ethyl ethanoate is made via route 1 starting with 200 g of ethanoic acid and excess ethanol to produce 190 g of ethyl ethanoate, what is the percentage yield? [3 marks]

6. 600 cm³ of a gaseous hydrocarbon was completely burned in oxygen and produced 1800 cm³ of carbon dioxide and 1800 cm³ of water vapour.

   a) Assuming all measurements are made at room temperature and pressure, calculate the number of moles of hydrocarbon that were burned. [2 marks]
   b) How many carbon atoms does each molecule of the hydrocarbon contain? [2 marks]
   c) A student suggested that the hydrocarbon was propane. Use evidence from the results of combustion to explain whether or not you agree. [4 marks]

7. When an airbag goes off in a car nitrogen gas is generated very quickly through the reaction below:

   $$2NaN_3(s) \longrightarrow 3N_2(g) + 2Na(s)$$

   In a typical airbag of 65 dm³ the mass of NaN_3 used is 130 g.

   a) Calculate the number of moles of NaN_3 in 130 g of NaN_3. [2 marks]
   b) Assuming that the reaction happens at room temperature 25°C and pressure 101 kPa, what volume of $N_2(g)$ would be generated? [1 mark]

   The $N_2(g)$ released fills the bag so that the actual volume is 65 dm³.

   c) Calculate the pressure generated inside the bag under these conditions. [4 marks]

DAY 4 — 60 Minutes

Chemical Equilibria

Dynamic equilibrium happens in a closed system when a reversible reaction has a rate of forward reaction equalling the rate of the backward reaction, i.e. products are being made at the same rate as they are being broken down.

$3H_2(g) + N_2(g) \rightleftharpoons 2NH_3(g)$ ammonia is made
$2NH_3(g) \rightleftharpoons 3H_2(g) + N_2(g)$ and broken down at the same rate

At equilibrium, the reaction vessel contains a mixture of reactants and products and there is no change in the concentration of either.

Equilibrium can only be reached in a closed system where neither reactants nor products can leave. If the products are removed then the reaction will continue to go in the forward direction.

If the forward reaction of an equilibrium is exothermic (ΔH –ve) then the reverse reaction is endothermic (ΔH +ve) by the same amount.

The **yield** is the percentage of the desired product in the equilibrium mixture.

Le Chatelier's Principle

Applies only to reactions that have reached equilibrium.

For a system at equilibrium, the concentration of reactants and products will change to oppose any change made to the system.

Changing the concentration of any reactant or product, or changing the pressure or temperature, unbalances the equilibrium. The rate of the forward and backward reactions changes so they are no longer going at the same rate. Eventually the system settles to equilibrium again, where both reactions are going at the same rate, but this is a different rate to that of the previous equilibrium. In the new equilibrium the concentration of reactants and products in the mixture is different. The factor that was changed (concentration, pressure or temperature) will be close to the value it had in the original equilibrium.

Other concentrations will change, using up added chemicals, restoring changed pressure or restoring changed temperature.

Adding Reactants Increases the Yield

e.g. adding H_2 to $3H_2(g) + N_2(g) \rightleftharpoons 2NH_3(g)$

The system **decreases** the concentration of H_2 by reacting it with the N_2 present in the equilibrium mixture making more NH_3. The concentrations of H_2, N_2 and NH_3 change until a new equilibrium is established.

In the new equilibrium mixture, the concentration of H_2 is close to its original concentration. The concentration of N_2 has decreased and the concentration of NH_3 has increased.

The position of equilibrium has moved to the right and the yield has increased.

Adding Products Decreases the Yield

e.g. adding NH_3 to $3H_2(g) + N_2(g) \rightleftharpoons 2NH_3(g)$

The system decreases the concentration of NH_3 by breaking it down to N_2 and H_2 and a new equilibrium is established.

In the new equilibrium the concentration of N_2 and H_2 are higher but the concentration of NH_3 is similar to the old equilibrium.

The position of equilibrium has moved to the left and the yield has decreased.

Increasing the Temperature Moves the Position of Equilibrium in the Endothermic Direction

$3H_2(g) + N_2(g) \rightleftharpoons 2NH_3(g)$ ΔH –92 kJ mol^{-1}

The negative enthalpy change means the reaction is exothermic in the forward direction.

If the temperature is increased, the system will favour the endothermic reaction, the concentration of reactants and products will change and this will reduce the temperature.

The temperature will be reduced if the rate of the endothermic reaction increases since this takes in heat. NH_3 is broken down at a faster rate until a new equilibrium is established.

In the new equilibrium, the concentration of NH_3 has decreased and the concentration of H_2 and N_2 have increased. The position of equilibrium has moved to the left and the yield has decreased.

Increasing temperature results in a higher concentration of the products of an equilibrium system where the forward reaction is endothermic. This means the yield increases. If the forward reaction is exothermic, increasing the temperature decreases the yield.

Decreasing the temperature has the opposite effect.

Increasing the Pressure Moves the Position of Equilibrium towards the Side with Fewer Moles of Gas

The concentration of reactants and products changes to reduce the pressure.

E.g. for $3H_2(g) + N_2(g) \rightleftharpoons 2NH_3(g)$

there are 4 moles of gas on the reactant side and 2 moles of gas on the product side.

All gases occupy the same volume and so exert the same pressure. The higher the percentage of products there are in the reaction mixture, the lower the pressure in the reaction vessel.

The system makes more NH_3 from N_2 and H_2. The position of equilibrium moves to the right and the yield increases.

Decreasing the pressure has the opposite effect.

If reactants and products have the same number of moles of gas then changing the pressure has no effect on the position of equilibrium.

Note: Increasing the pressure also increases the **rate** of both forward and backward reaction for a system that has gases in it.

Adding a Catalyst Does Not Affect the Yield

A catalyst speeds up both the forward and backward reaction to the same extent so has no effect on the position of equilibrium. It does speed up the rate at which equilibrium is reached, or it allows equilibrium to be reached in a reasonable time at a lower temperature.

Equilibrium Constant K_c

- Shows the position of equilibrium for a reaction.
- Large K_c (>1) means the position of equilibrium is towards the products.
- Small K_c (<1) means the equilibrium is towards the reactants.

The Equilibrium Expression

The equation to find the equilibrium constant is written from the reaction equation:

$$3H_2(g) + N_2(g) \rightleftharpoons 2NH_3(g) \qquad K_C = \frac{[NH_3]^2}{[H_2]^3 [N_2]}$$

Each reactant and product is written as a concentration by using square brackets.	$[NH_3(g)]$ $[H_2(g)]$ $[N_2(g)]$
Each concentration is raised to the power of its coefficient in the reaction equation.	$2NH_3(g) = [NH_3(g)]^2$ $3H_2(g) = [H_2(g)]^3$ $N_2(g) = [N_2(g)]$
The products are multiplied together on the top of the equation and the reactants multiplied together on the bottom of the equation.	$\frac{[NH_3(g)]^2}{[H_2(g)]^3 [N_2(g)]}$

Calculating the Quantity of a Substance at Equilibrium

The concentration of any substance in the equilibrium mixture is found by adding together the initial concentration and the change in concentration for that substance.

Find the starting moles of all substances in the equation (often 0 for one side of the reaction):

$$CO(g) + 2H_2(g) \rightleftharpoons CH_3OH(g)$$

If the starting quantity of $CO(g)$ was 1 mol and the starting quantity of $H_2(g)$ was 2 mol and the equilibrium concentration of $CO(g)$ was 0.3 mol dm^{-3}:

DAY 4

1. Calculate how many moles of CO have been converted to products. Starting moles – equilibrium moles.

   1 mole – 0.3 moles = 0.7 moles

2. Calculate how many moles of products have been made.

   Every 1 mole of CO produces 1 mole CH_3OH.
   Moles of CH_3OH = 0.7

3. Calculate how many moles of H_2 have been used up.

   Every 1 mole of CH_3OH uses 2 moles of H_2.
   Moles of H_2 used = 0.7 × 2 = 1.4

4. Calculate how many moles of H_2 remain at equilibrium. Starting moles – moles used.

   Moles of H_2 at equilibrium = 2 – 1.4 = 0.6

This is easily summarised in a table.

	CO	H_2	CH_3OH
starting moles	1	2	0
change in moles	1 – 0.3 = 0.7	–(2 × 0.7) = –1.4	0 + 0.7 = 0.7
equilibrium moles	0.3	2 – 1.4 = 0.6	0.7

The concentration of each substance at equilibrium can be calculated provided the volume of the mixture is known. This allows the equilibrium constant to be calculated.

Equilibria in Industry

Very high temperatures are costly as they use a lot of energy. Often, to make the energy, fossil fuels are burned and this releases carbon dioxide, which has been linked to climate change.

Making equipment that can withstand high pressures is costly. A compressor required to maintain high pressure has high energy costs.

Industrial chemists must compromise between reaction conditions that give a high yield, with a suitable rate of reaction and the financial and environmental costs of achieving them.

A catalyst is always a good way of achieving equilibrium faster with lower energy costs.

Exothermic reactions give the highest equilibrium yield when the temperature is low. Low temperatures result in a slow rate of reaction. It may be more economical to accept a lower yield that can be achieved more quickly.

The Haber process ideally requires low temperatures and high pressures. In industry a temperature of 450°C and pressure of 200 atmospheres is used as a compromise.

QUICK TEST

1. State Le Chatelier's principle.

2. $CO_2(g) \rightleftharpoons CO_2(aq)$
   What is the effect on the position of equilibrium of increasing the concentration of $CO_2(g)$ in this system?

3. $H_2(g) + I_2(g) \rightleftharpoons 2HI(g)$ ΔH −9.6 kJmol^{-1}
   Describe what will happen to the position of equilibrium if the temperature is increased.

4. What will happen to the position of equilibrium if the pressure is reduced?

5. Describe the effect on the position of equilibrium if a catalyst is added.

6. Explain why industrial chemists sometimes choose reaction conditions that do not give the best equilibrium yield of product.

PRACTICE QUESTIONS

1. Nitrogen and oxygen react together to produce nitrogen oxide.

$$N_2(g) + O_2(g) \rightleftharpoons 2NO(g)$$

Which will cause the position of equilibrium to move to the right? [1 mark]

- **A** Increasing the pressure
- **B** Adding a catalyst
- **C** Adding more nitrogen
- **D** Decreasing the pressure

2. Sulfur dioxide reacts with water.

$$SO_2(g) + H_2O(g) \rightleftharpoons H^+(aq) + HSO_3^-(aq)$$

What would be the effect of adding more water vapour to the equilibrium mixture? [1 mark]

- **A** The concentration of $H^+(aq)$ would decrease.
- **B** The concentration of $SO_2(g)$ would increase.
- **C** There will be no change in the quantity of $HSO_3^-(aq)$.
- **D** The quantity of $SO_2(g)$ would decrease.

3. Yellow chromate ions react together with acids to form red dichromate ions and water.

$$2CrO_4^{2-}(aq) + 2H^+(aq) \rightleftharpoons Cr_2O_7^{2-}(aq) + H_2O(l)$$

a) How would the colour of the mixture change if $H^+(aq)$ was added? [1 mark]

b) What could be added to the equilibrium to encourage a yellow colour? [1 mark]

4. Hydrogen can be generated by heating together methane with carbon monoxide in a process known as steam reforming.

$$CH_4(g) + H_2O(g) \rightleftharpoons CO(g) + 2H_2(g) \quad \Delta H +210 \text{ kJ mol}^{-1}$$

a) Explain why this reaction will never result in 100% yield of hydrogen. [2 marks]

b) Justify which conditions of pressure would give the best yield of hydrogen. [2 marks]

c) Explain why the addition of a nickel catalyst makes the reaction more economical. [2 marks]

The carbon monoxide produced in the reaction can be treated further to release more hydrogen.

$$CO(g) + H_2O(g) \rightleftharpoons CO_2(g) + H_2(g) \quad \Delta H -42 \text{ kJ mol}^{-1}$$

d) This reaction is carried out at a lower temperature to increase the yield. Explain why the reaction with methane and water vapour is carried out at a higher temperature than the reaction with carbon monoxide and water vapour. [3 marks]

e) The equilibrium constant for steam reforming at 700°C is larger than 100 mol² dm⁻⁶. What does this value suggest about the position of equilibrium? [1 mark]

f) Write the equilibrium expression for steam reforming that could be used to calculate the equilibrium constant. [3 marks]

DAY 4 — 60 Minutes

Redox

Oxidation of an element can be described as:
- addition of oxygen
- loss of hydrogen
- loss of electrons
- increase in oxidation number.

Reduction can be described as:
- loss of oxygen
- addition of hydrogen
- gain of electrons
- decrease in oxidation number.

Oxidation
Is
Loss of electrons
Reduction
Is
Gain of electrons

In a chemical reaction, oxidation and reduction always go together. If one substance is oxidised, another has been reduced.

The oxidising agent in a redox reaction is the reactant that has been reduced; it has <u>received the electrons</u>.

The reducing agent is the reactant that has been oxidised; it has <u>donated the electrons</u>.

Oxidation Number

Oxidation numbers are written sign first, number second, e.g. oxidation number for Mg^{2+} is +2.

- The oxidation number of an element is always zero.
- The oxidation number for a monoatomic ion is the charge on the ion.
- The sum of the oxidation numbers in a neutral compound is zero.
- The sum of the oxidation numbers in a compound ion is the charge on the ion.

Oxidation numbers for elements in a compound can be calculated using a set of rules.

The oxidation number of	Exceptions
fluorine is always –1	None
oxygen is –2	When combined with fluorine. In peroxides* O is –1 e.g. H_2O_2
hydrogen is +1	In metal hydrides e.g. in NaH H is –1

* compounds containing O–O bonds

Finding the Oxidation Number of a Compound

Examples
1. The oxidation number of magnesium and chlorine in $MgCl_2$:
   The oxidation number is the same as the charge on the ion.
   Magnesium is +2   Chlorine is –1

2. The oxidation number of nitrogen in NO_2:
   The oxidation number of oxygen is –2
   The sum of the oxidation number for a compound = 0
   Two oxygens = 2 × –2 = –4
   –4 + oxidation number of nitrogen = 0
   Oxidation number of nitrogen = +4

3. The oxidation number of sulfur in SO_4^{2-}:
   The oxidation number of oxygen is –2.
   The sum of the oxidation numbers = charge on the ion = –2
   (4 × –2) + oxidation number of S = –2
   Oxidation number of S = +6

4. The oxidation number of Cr in $K_2Cr_2O_7$:
   The oxidation number of a monoatomic ion is the charge on the ion: K = +1
   The sum of the oxidation numbers in a compound = 0
   (2 × +1) = sum of oxidation numbers in Cr_2O_7 = –2

> The sum of the oxidation number for a compound ion = charge on the ion
> Charge on $Cr_2O_7 = 2-$
> The oxidation number of oxygen is -2.
> $(7 \times -2) + 2 \times$ oxidation number of $Cr = -2$
> $-14 + 2 = 2 \times$ oxidation number of Cr
> Oxidation number of $Cr = +6$

Writing Half Equations

Redox reactions can be written as full equations, ionic equations, or separated into two half equations. One represents what happens to the oxidising agent and the other what happens to the reducing agent.

$$Mg + Cl_2 \longrightarrow MgCl_2$$

Oxidation half equation $\quad Mg \longrightarrow Mg^{2+} + 2e^-$

Reduction half equation $\quad Cl_2 + 2e^- \longrightarrow 2Cl^-$

Identifying a Redox Reaction

Write out the reaction equation and work out the oxidation number of each element in the reactants and products.

$$2NaBr + 2H_2SO_4 \longrightarrow 2NaHSO_4 + SO_2 + H_2O + Br_2$$

Na	+1		+1	
Br	−1			0
H		+1	+1	+1
S		+6		+4

Look for changes in oxidation number.

Disproportionation Reactions

The same element is both oxidised and reduced in the same chemical reaction.

$$Cl_2 + H_2O \longrightarrow HCl + HClO$$

Cl $\quad$ 0 $\quad\quad\quad\quad$ −1 $\quad$ +1

Chlorine is oxidised $\quad 0(Cl_2) \longrightarrow +1 (HClO)$
and reduced $\quad\quad\quad 0(Cl_2) \longrightarrow -1 (Cl^-)$.

Naming Compounds Containing Elements with Variable Oxidation States

The name of the compound shows the oxidation number of the element using Roman numerals in brackets after the symbol for the element.

iron(II) chloride $FeCl_2$ $\quad$ iron(III) chloride $FeCl_3$

Compound ions show the oxidation number of **the element that is not oxygen** in Roman numerals.

nitrate(V) NO_3^- $\quad\quad\quad$ nitrate(III) NO_2^-

chlorate(V) ClO_3^- $\quad\quad\quad$ chlorate(I) ClO^-

Working Out the Formula of a Compound from the Name — neat

What is the formula of potassium chlorate(III)?
Write the elements in the compound: K Cl O

Write the known oxidation numbers: +1 +3 −2

The overall charge on a compound is 0.
Since potassium is +1, chlorate = −1
The sum of the oxidation numbers for a compound ion is the same as the charge on the ion.
Sum of oxidation numbers of chlorine + oxygen = −1
If n = number of atoms of oxygen in the compound:

$+3 + n(-2) = +1$ $\quad -1 - 3 = n(-2)$ $\quad -4 = n(-2)$
$\frac{-4}{-2} = n$ $\quad\quad n = 2$

Formula of potassium chlorate(III) = $KClO_2$

Note: The names nitrate and sulfate with no oxidation state given should be considered to be nitrate(V) NO_3^- and sulfate(VI) SO_4^{2-}.

Redox in Electrolysis (OCR B only)

If an electric current is passed through a liquid containing ions then the ions will migrate towards the electrodes.

Reduction at the Cathode

Positive ions (cations) migrate towards the cathode (negative electrode). At the cathode they accept electrons and are reduced to atoms (discharged).

$$M^{n+} + ne^- \longrightarrow M$$

Oxidation at the Anode

Negative ions (anions) migrate towards the anode (positive electrode). At the anode they lose electrons and are oxidised to atoms (discharged).

$$A^{n-} \longrightarrow A + ne^-$$

DAY 4

Electrolysis of Molten Salts

e.g. NaCl(l)

Cathode $Na^+ + e^- \longrightarrow Na$

Anode $2Cl^- \longrightarrow Cl_2 + 2e^-$

Electrolysis of Aqueous Solutions

Aqueous solutions contain the dissolved cations and anions and also H^+ and OH^- from water.

H^+ can be reduced at the cathode.
$$2H^+ + 2e^- \longrightarrow H_2$$

OH^- can be oxidised at the anode.
$$4OH^- \longrightarrow O_2 + 2H_2O + 4e^-$$

For electrolysis in aqueous solution there is competition between the metal ions in solution and the hydrogen ions. If the metal is above hydrogen in the metal reactivity series then hydrogen is discharged rather than the metal, e.g. aqueous Group 1 and 2 metals and aluminium salts discharge hydrogen.

At the anode, anions are discharged. If halide ions are present in solution these are discharged. If not, oxygen is formed from the water.

Example
Electrolysis of sodium chloride solution

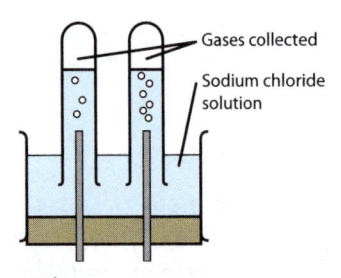

Cathode reaction
$$2H^+ + 2e^- \longrightarrow H_2$$

Anode reaction
$$2Cl^- \longrightarrow Cl_2 + 2e^-$$

Electrolysis with Reactive Electrodes

If reactive electrodes are used, the anode atoms lose electrons to the circuit leaving positive ions. These dissolve into the electrolysis solution and begin to migrate towards the cathode.

Example
Electrolysis of copper sulfate solution using copper electrodes

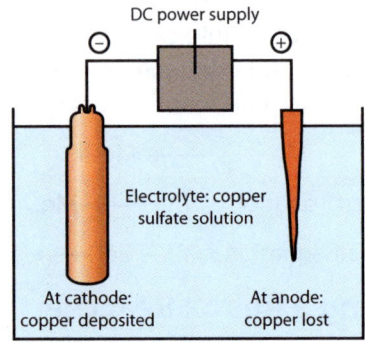

Electrolysis using copper electrodes.

The metal copper anode becomes copper ions in solution.
$$Cu \longrightarrow Cu^{2+} + 2e^-$$

The cathode gains copper.
$$Cu^{2+} + 2e^- \longrightarrow Cu$$

QUICK TEST

1. In $C + 2H_2 \longrightarrow CH_4$ is carbon oxidised or reduced?

2. Which element is oxidised when copper and sulfur react together?

3. $Fe^{2+} \longrightarrow Fe^{3+} + e^-$
   Classify this half equation as oxidation or reduction.

4. State the name of the ClO_3^- ion.

5. State the formula of the nitrate(III) ion.

6. List the chemicals discharged at each electrode in the electrolysis of $H_2SO_4(aq)$.

PRACTICE QUESTIONS

1. In the reaction

$$Cr_2O_7^{2-} + 6Fe^{2+} + 14H^+ \longrightarrow 3Cr^{3+} + 6Fe^{3+} + 7H_2O$$

Fe^{2+} is [1 mark]

- **A** the oxidising agent
- **B** oxidised to Fe^{2+}
- **C** the reducing agent
- **D** receiving electrons from Cr.

2. The oxidation number of Cr in $K_2Cr_2O_7$ is [1 mark]

- **A** +12
- **B** +2
- **C** +6
- **D** −6

3. Which of the following is not a redox reaction? [1 mark]

- **A** $2Mg(s) + O_2(g) \longrightarrow 2MgO(s)$
- **B** $H_2SO_4(aq) + Na_2CO_3(aq) \longrightarrow Na_2SO_4(aq) + CO_2(g) + H_2O(l)$
- **C** $Zn(s) + 2HCl(aq) \longrightarrow ZnCl_2(aq) + H_2O(l)$
- **D** $Br_2(aq) + 2KI(aq) \longrightarrow 2KBr(aq) + I_2(aq)$

4. During the electrolysis of NaCl(l) the following ions are discharged: [1 mark]

- **A** Na^+ and Cl^-
- **B** H^+ and Cl^-
- **C** Na^+ and OH^-
- **D** H^+ and OH^-

5. Bromine is present in high concentration in water from the Dead Sea. The first step in extraction of bromine from seawater is to bubble chlorine through it. This displaces the bromine, which is then concentrated and purified.

- **a)** Write an equation for the displacement reaction of bromine from seawater. [1 mark]
- **b)** What is the reducing agent in this reaction? [1 mark]

One use for the bromine produced is in the manufacture of silver bromide for photographic film. The silver bromide is produce by reacting silver nitrate with potassium bromide.

$$AgNO_3(aq) + KBr(aq) \longrightarrow AgBr(s) + KNO_3(aq)$$

The AgBr formed is light sensitive and the bromide ion loses an electron when the light strikes it.

$$2AgBr(s) \longrightarrow 2Ag(s) + Br_2(l)$$

- **c)** Show that the reaction of bromide ions with silver nitrate is not a redox reaction. [2 marks]
- **d)** What happens to the electron that is lost by bromide when light strikes it? [1 mark]

6. $TiCl_4$ can be prepared by thermal disproportionation of $TiCl_3$ in the following reaction:

$$2TiCl_3 \longrightarrow TiCl_2 + TiCl_4$$

- **a)** What name is given to $TiCl_4$? [1 mark]
- **b)** Explain what is meant by disproportionation and show that this reaction is an example of this type of reaction. [3 marks]

63

DAY 4 — 60 Minutes

Nuclear Reactions & Radiation

(OCR B and WJEC only)

Chemical reactions involve changes in the electrons of atoms. The number of atoms of each element never changes although some atoms may become ions and vice versa.

Nuclear reactions involve changes in the particles *(nucleons)* in the nucleus. This may result in the formation of different elements.

Radioactive Decay

Isotopes of some elements have unstable nuclei that break down spontaneously and emit particles and radiation. These are known as radioisotopes.

α-decay

Particle emitted	Helium nucleus $^{4}_{2}\text{He}^{2+}$
Penetrating power	A few centimetres in air. Stopped by paper
Deflection in electric field	Weak, towards the negative plate
Deflection in a magnetic field	Weak and showing a positive charge
Effect on nucleus	Loss of 2 protons and 2 neutrons
Effect on element name	Moves back two places in the periodic table

β-decay

Particle emitted	Electron $^{0}_{-1}\text{e}$
Effect on nucleus	Loss of 1 neutron; gain of 1 proton *— wild —*
Penetrating power	A few metres in air. Stopped by aluminium foil
Deflection in an electric field	Strong, towards the positive plate
Deflection in a magnetic field	Strong and showing a negative charge
Effect on element name	Moves forward one place in the periodic table

Positron Emission / β⁺-decay

Particle emitted	Positron $^{0}_{+1}\text{e}$ (anti-electron)
Effect on nucleus	Loss of 1 proton; gain of 1 neutron
Effect on element name	Moves back one place in the periodic table

Electron Capture

Particle emitted	None: one electron from the innermost shell is absorbed into the nucleus
Effect on nucleus	Loss of 1 proton; gain of 1 neutron
Effect on element name	Moves back one place in the periodic table

γ-decay

Particle emitted	Electromagnetic radiation, often emitted along with other radioactive decay
Penetrating power	Long distances *hella strong*
Deflection in electric and magnetic fields	None
Effect on nucleus	No change in particles
Effect on element name	None

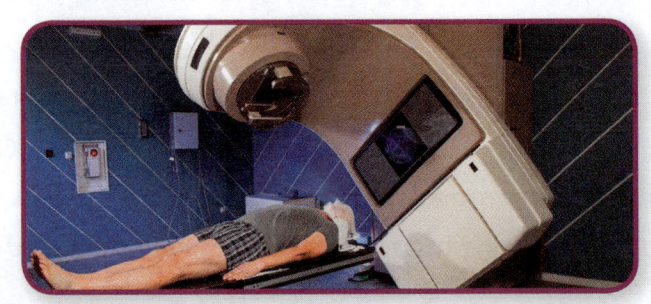

why is an α weak in a B-field when it has +2 charge, and β strong w/ only −1 charge?

64

Nuclear Equations

Show the mass number and proton number after radioactive decay or nuclear fusion.

α-decay of uranium

$$^{238}_{92}U \longrightarrow {}^{234}_{90}Th + {}^{4}_{2}He$$

β-decay of carbon-14

$$^{14}_{6}C \longrightarrow {}^{14}_{7}N + {}^{0}_{-1}e$$

Positron emission of carbon-11

$$^{11}_{6}C \longrightarrow {}^{11}_{5}B + {}^{0}_{+1}e$$

Electron capture of aluminium-26

$$^{26}_{13}Al + {}^{0}_{-1}e \longrightarrow {}^{26}_{12}Mg$$

Many radioisotopes decay into another radioisotope, which in turn decays into another radioisotope. Eventually a stable isotope is produced, which ends the decay chain.

$$^{210}_{82}Pb \longrightarrow {}^{210}_{83}Bi \longrightarrow {}^{210}_{84}Po \longrightarrow {}^{206}_{82}Pb$$

Uses of Radioactive Decay

Medical

Ionising radiation is damaging to living cells. It can result in cell death or damage to DNA.

A β emitter can be used to destroy tumours, e.g. iodine-131 is used for thyroid cancer.

Since the radiation can be detected outside the body, radioisotopes can be placed in molecules and then ingested (eaten). The movement of the radioisotope can be traced externally. This allows detection of tumours and a better understanding of the metabolism of individuals and of drugs.

Industrial

Radioisotopes can be used to trace leakages in underground pipes.

Firing β radiation at metal sheets and then measuring the extent of penetration to the far side of the sheet allows the thickness of the material to be judged.

Radioactive Half-lives

The half-life is the time taken for one half of the radioactive nuclei to decay. The half-life of an isotope is not affected by chemical reactions.

Using Half-lives to Calculate the Time Required for Decay to Decrease

The half-life of iodine-131 is 8.1 days.

If the starting radioactive count of a sample was 256, how long would it take for the sample to decay to below 20 counts?

Half-life	Time (days)	Count
first	8.1	128
second	16.2	64
third	24.3	32
fourth	32.4	16

Using Radioactive Isotopes to Calculate Time Passed

Some rocks contain radioisotopes that decay slowly into a stable isotope. The proportion of the original isotope to the stable isotope reflects how many half-lives have passed. This gives an estimate of the age of the rock.

A problem in radioactive dating of rock is when there is loss or addition of radioactive material to the rock over geological time, e.g. if one member of the decay chain is a gas such as radon. *Dynamic Earth*

Radiocarbon Dating

Living organisms take in food containing carbon-12 and carbon-14 in a constant ratio. The half-life of carbon-14 is 5730 years.

Once the organism dies the carbon-14 continues to decay and is no longer replaced, so the carbon-14 : carbon-12 ratio changes. This can be used to estimate the time since the organism died.

Problems with radiocarbon dating include:
- contamination with modern materials. Since the levels of radioactivity are often very small, a small contamination can have a large effect on the apparent date.
- the uneven distribution of carbon-14 in the atmosphere with time and geographical location means the presumed starting ratio may be incorrect.

DAY 4

Fusion Reactions

If the nuclei of elements are made to collide with very high energy they may fuse together to form a new heavier element. This is how the heavier elements are thought to have formed in the centre of stars.

$$_1^1H + {}_1^1H \longrightarrow {}_1^2H + {}_{+1}^0e$$

$$_1^1H + {}_1^2H \longrightarrow {}_2^3He$$

Electromagnetic Radiation and Atoms

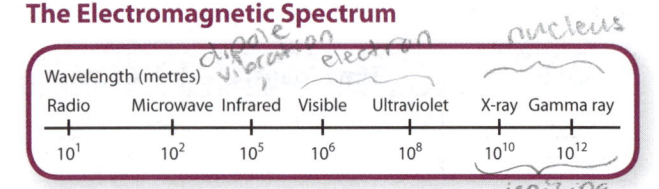

The Electromagnetic Spectrum

Wavelength (metres)						
Radio	Microwave	Infrared	Visible	Ultraviolet	X-ray	Gamma ray
10^1	10^2	10^5	10^6	10^8	10^{10}	10^{12}

Electromagnetic radiation can be thought of as a stream of photons. The energy of the photons depends on the frequency of the radiation.

Energy = Planck's constant × frequency
$E = hf$ or $E = hv$

Wavelength and frequency are related by the speed of light.

Speed of light = frequency × wavelength
$c = f\lambda$ or $c = v\lambda$

Electrons are placed within fixed energy positions around the nucleus. To transfer from a lower energy position to a higher one they must absorb exactly the energy difference (ΔE) between the two positions. To fall from a higher energy position to a lower one they must release exactly the energy difference between the positions.

The energy of photons in the visible and UV region of the electromagnetic spectrum is of the same order as the energy differences between electron energy levels.

If electrons absorb a photon of the correct energy they can move to a higher energy position. When they fall to a lower energy position they release a photon of specific energy.

Since $E = hf$ the absorption or emission is seen as a specific frequency of light.

The Lyman series shows the frequencies of light emitted from a hydrogen atom as electrons fall from higher to lower energy levels.

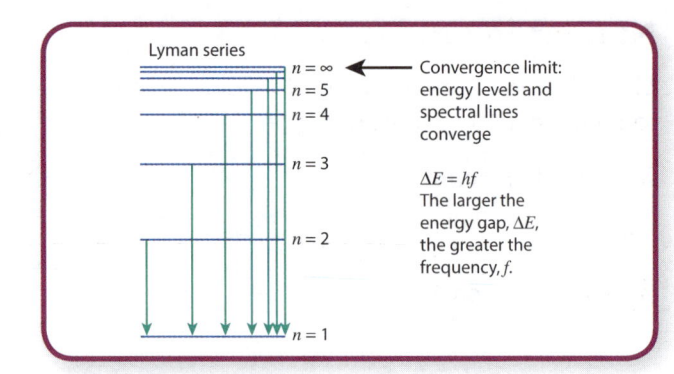

Lyman series

Convergence limit: energy levels and spectral lines converge

$\Delta E = hf$
The larger the energy gap, ΔE, the greater the frequency, f.

As the energy increases the lines get closer together until eventually they converge. This is the minimum energy required for the electron to escape the atom, the ionisation energy.

The electron energy levels are unique for each element so each element has a unique emission spectrum that can be used to identify it. The spectra appear as coloured lines on a dark background.

Photons of correct energy, hence correct frequency of light, can be absorbed to raise electrons to higher energy levels. If white light is shone on the element, these absorbed photons will be seen as dark lines on the rainbow of the spectrum.

QUICK TEST

1. Which element results from the α decay of uranium?
2. State which particle has been emitted from the following nuclear reaction: $_6^{11}C \longrightarrow {}_5^{11}B + ?$
3. Suggest what should be used to stop β particles from penetrating.
4. What is the energy of photons with frequency 4.5×10^{14}Hz (Planck's constant = 6.63×10^{-34}J Hz^{-1})
5. Iodine-131 has a half-life of 8 days. Calculate how long would it take for a count of 120 to reduce to 30.
6. Describe the appearance of an emission spectrum.

PRACTICE QUESTIONS

1. What is the change in the atom following electron capture? [1 mark]
   - A The element becomes a positive ion. ☐
   - B The atomic number decreases by one. ☐
   - C The atomic number increases by one. ☐
   - D The nucleus is unchanged. ☐

2. Which of the following are properties of a β particle? [1 mark]
   - A It is not deflected by a magnetic field. ☐
   - B It is stopped by paper. ☐
   - C It is deflected to the positive electric plate. ☐
   - D It does not affect the nucleus. ☐

3. One type of radioactive decay results in the emission of a positron. The effect on the nucleus is [1 mark]
   - A an increase in nuclear charge ☐
   - B the change of a proton into a neutron ☐
   - C a loss in atomic mass ☐
   - D a gain in atomic mass. ☐

4. Polonium-210 is an α emitter with a half-life of 138 days.
   - a) What is the decay product of polonium-210? [1 mark]
   - b) Polonium-210 is very dangerous if ingested. Explain why. [2 marks]
   - c) What is meant by the half-life of a radioisotope? [1 mark]
   - d) How long would it take a sample of polonium with a radioactive count of 800 to reduce to 100? [1 mark]
   - e) Radioisotopes are sometimes used as biological tracers. Why would an isotope that undergoes β-decay be chosen as a tracer? [1 mark]

5. Scientists are working towards producing energy from nuclear fusion. One likely reaction is the fusion of tritium and deuterium to form helium plus a neutron. Deuterium is an isotope of hydrogen with one neutron in the nucleus while tritium is hydrogen with two neutrons in the nucleus.
   - a) Complete the nuclear equation for this reaction.

     ____ + ____ ⟶ $^{4}_{2}$He + n [3 marks]
   - b) It is thought that fusion reactions in the centre of stars have resulted in the heavier elements of the periodic table. Suggest what element might be produced by the nuclear fusion of two $^{4}_{2}$He atoms. [1 mark]

6. Chlorine radicals are produced in the stratosphere when C—Cl bonds absorb ultraviolet light. The bond enthalpy of the C—Cl bond is +346 kJ mol^{-1}. The energy of a photon is given by $E = h\nu$. The value of $h = 6.63 \times 10^{-34}$ J Hz^{-1}.
   - a) How much energy is needed to break one C—Cl bond? [2 marks]
   - b) What frequency of light would provide this energy? [2 marks]

DAY 5 60 Minutes

Periodicity/Group 2 Elements

The Periodic Table

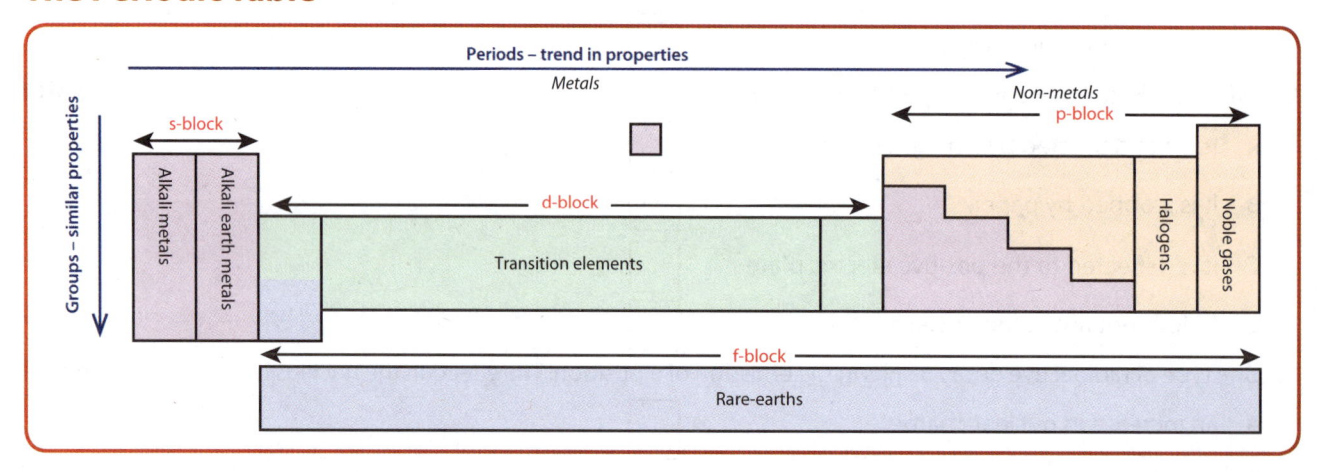

The modern periodic table lists elements in order of their proton number and places elements with the same number of outer shell electrons in the same column. A row is called a period and a column is called a group. Elements in the same period have the same number of electron shells.

The table is divided into blocks: s, p, d and f.

Elements in the same block have their highest energy electron in the same type of sub-shell.

Example
p-block   Si [Ne] $3s^2$ $3p^2$ Br [Ar] $3d^{10}$ $4s^2$ $4p^5$
d-block   V [Ar] $3d^3$ $4s^2$ Zn [Ar] $3d^{10}$ $4s^2$

Trends Down a Group

For each subsequent member going down the group:
- atomic radius increases because there is one additional shell of electrons
- The total number of protons – number of shielding electrons in inner shells is the same for all members, e.g. for Group 1
Li: +3 – 2 = 1,
Cs: +55 – 54 = 1
- the first ionisation energy decreases because the outer shell electrons are further from the positive pull of the nucleus so are held less strongly to the atom.

Melting point for Groups 1 and 2 (all metals) generally decreases. Increased ionic radius means weaker attraction to the delocalised electrons in the metallic lattice so less energy is needed to break them. (Differences in structure mean the general pattern is not consistent.)

Melting points and boiling points for Group 7 and Group 0 (this group are all gases) increase down the group as the increased number of electrons means increased strength of van der Waals forces so more energy is needed to break them.

Other groups cross the metal/non-metal divide and show more complex trends.

Trends Across a Period

Each period shows a similar pattern, hence properties recur periodically.

For each period, the Group 1 element has the highest atomic radius and lowest first ionisation energy while the Group 0 element has the smallest atomic radius and highest first ionisation energy.

Metal elements are on the left of the table and non-metals on the right. **Electrical conductivity increases** across the metallic elements as the number of electrons donated to the mobile sea of electrons increases. It then drops to zero for the non-metals.

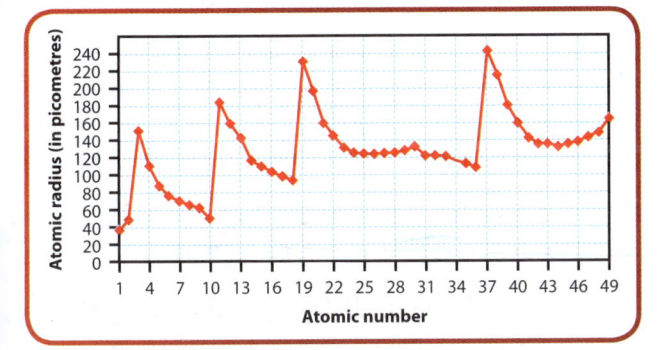

Atomic Radius of Elements vs Atomic Number

Atomic radius decreases across a period because the nuclear charge increases (one more proton per element). The outer electron shell is more strongly attracted to the increasingly positive nucleus and is pulled inwards, reducing the radius.

First ionisation energy increases across a period because nuclear charge increases, the outer shell electrons are closer to the nucleus and shielding is almost constant. This means more energy is required to remove an electron. (For a more detailed review of first ionisation energy see page 13 Electron Configuration.)

Electronegativity increases across a period for the same reasons as the increase in first ionisation energy. The increasing nuclear charge is increasingly able to attract electrons in a covalent bond.

Trends in Period 2 and 3

Period 2	Li	Be	B	C	N	O	F	Ne
Structure	giant metallic		giant covalent		simple covalent (diatomic)			atomic

Period 3	Na	Mg	Al	Si	P	S	Cl	Ar
Structure	giant metallic			giant covalent	P_4	S_8	Cl_2	atomic
					simple covalent			

Melting Point

Trends in melting point in Periods 2 and 3 can be explained by the changes in structure across the period.

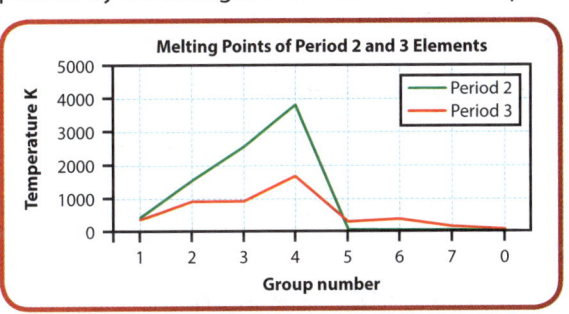

Melting point increases across the period for the metals. The charge on the ion in the metallic lattice increases from Na^+ to Mg^{2+} to Al^{3+} while the ionic radius decreases. This means there is a stronger attraction between the ions and the delocalised electrons in the giant metallic structure (see page 28). More energy is needed to allow them to move past each other.

Melting point peaks at the giant covalent structures. Many strong covalent bonds hold each atom in position and these must be broken to allow the atoms to move past each other. This requires a large amount of energy (see page 26).

Melting point is low for the simple covalent structures. Simple molecules are held together by weak instantaneous dipole–induced dipole forces, which require little energy to break (see page 25). The melting point depends on the number of electrons and this is often determined by the size of the molecules (see table on the left).

Group 2 Elements

Group 2 elements are metals with two electrons in their outer shell. They react to lose two electrons to form a 2^+ ion.

$$M \longrightarrow M^{2+} + 2e^- \quad \text{oxidation}$$

Changes Going Down the Group

Be
Mg
Ca
Sr
Ba

Atomic radius and ionic radius increase as each subsequent element has one additional shell of electrons.

The two outer shell electrons are further from the positive pull of the nucleus and so are more easily lost. Reactivity increases going down the group.

DAY 5

Charge Density of M²⁺ Ions
Decreases down the group as the charge of the ion remains the same but radius increases.

Reactions of Group 2 Elements

Group 2 element + oxygen forms an oxide.

$$2M(s) + O_2(g) \longrightarrow 2MO(s)$$

Group 2 element + water generally forms a hydroxide + hydrogen.

$$M(s) + 2H_2O(l) \longrightarrow M(OH)_2(s) + H_2(g)$$

Ca to Ba react with cold water. Mg reacts slowly with water and reacts quickly with steam to form MgO and hydrogen. Be does not react with water.

Group 2 metal + acid forms a salt + hydrogen.

$$M(s) + 2H^+(aq) \longrightarrow M^{2+}(aq) + H_2(g)$$

$$Mg(s) + 2HCl(aq) \longrightarrow MgCl_2(aq) + H_2(g)$$

Note: Reaction with H_2SO_4 gives an insoluble sulfate with Ca, Sr and Ba that coats the metal and prevents further reaction. Reaction with more concentrated nitric acid reduces the nitrate group to nitrogen oxides.

Group 2 element + chlorine forms metal chlorides.

$$M(s) + Cl_2(g) \longrightarrow MCl_2(s)$$

Reactions of Group 2 Compounds

Group 2 hydroxide + dilute acid give a salt + water.

$$M(OH)_2(s) + 2H^+(aq) \longrightarrow M^{2+}(aq) + 2H_2O(l)$$

e.g $Sr(OH)_2(s) + 2HCl(aq) \longrightarrow SrCl_2(aq) + 2H_2O(l)$

Group 2 oxide + water forms a hydroxide.

$$MO(s) + H_2O(l) \longrightarrow M(OH)_2(s)$$

Solubility of Group 2 Hydroxides
Solubility increases down the group. $Mg(OH)_2$ releases fewer OH⁻ ions into solution than $Sr(OH)_2$ so $Mg(OH)_2(aq)$ has a lower pH than $Sr(OH)_2(aq)$.

Solubility of Group 2 Sulfates
Solubility decreases down the group. $BaSO_4$ is insoluble.

Thermal Decomposition of the Group 2 Carbonates

$$MCO_3(s) \longrightarrow MO(s) + CO_2(g)$$

Thermal stability increases going down the group. At the top of the group the high charge density of the cation causes distortion in the CO_3^{2-} ion. This polarises it making it easier to break down and release CO_2.

Thermal decomposition of Group 2 nitrates gives metal oxides, nitrogen(IV) oxide and oxygen.

$$2M(NO_3)_2(s) \longrightarrow 2MO(s) + 4NO_2(g) + O_2(g)$$

Thermal stability increases down the group as the polarising ability of the cation decreases.

Thermal Stability of Group 1 Compounds (Edexcel only)

Group 1 metal ions have a lower charge density than Group 2 ions. This makes the carbonates very thermally stable by comparison to those of Group 2.

The nitrates of Group 1 decompose to form nitrate(III) salts and oxygen.

$$2MNO_3(s) \longrightarrow 2MNO_2(s) + O_2(g)$$

Exceptions: $LiNO_3$ which behaves as Group 2 metal nitrates.

QUICK TEST

1. Name a p-block element.
2. Explain why Fe is called a d-block element.
3. Explain why the first ionisation energy of barium is lower than that of magnesium.
4. Explain why S has a higher melting point than P.
5. Describe the trend in pH of the Group 2 hydroxides.
6. Write an equation for the thermal decomposition of $Mg(NO_3)_2$.

PRACTICE QUESTIONS

1. Which of the following is both a p-block element and has the highest ionisation energy in its group? **[1 mark]**

   A Li ☐   B Si ☐   C O ☐   D I ☐

2. Identify the false statement. **[1 mark]**

   A Group 2 nitrates decompose to produce a brown gas. ☐

   B The gas released when Group 2 carbonates decompose turns limewater milky. ☐

   C The gas released when a Group 2 nitrate decomposes will relight a glowing splint. ☐

   D The gas released when a Group 2 metal reacts with acid will put out a lighted splint. ☐

3. This graph shows the trend in melting points of elements in Groups 2 and 3.

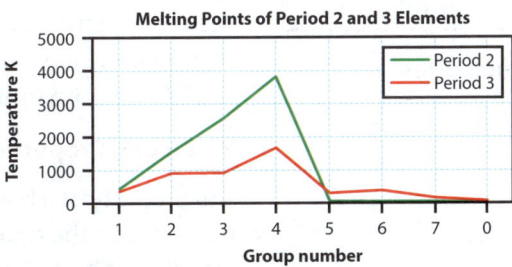

   a) Explain why the highest melting point for both periods occurs at the Group 4 element. **[3 marks]**

   b) Use ideas about structure and bonding to give reasons for the increase in melting point from Group 1 to Group 3 in Period 2. **[4 marks]**

   c) The metal elements in Period 2 have higher melting temperatures than those in Period 3. Suggest a reason for this difference. **[2 marks]**

   d) In Period 2 the melting points of the elements in Groups 5–7 are very similar. In Period 3 the melting point of the element in Group 6 is higher than those in Groups 5 and 7. Name the elements and give reasons for the differences in melting point. **[5 marks]**

4. A student carried out an experiment to compare the thermal stability of $BaCO_3$, $MgCO_3$ and $CaCO_3$.

   a) Draw a labelled diagram of suitable apparatus for this experiment. **[2 marks]**

   b) Describe a suitable procedure. **[2 marks]**

   c) Predict the outcome of this experiment. **[1 mark]**

   d) Give an explanation for your prediction. **[3 marks]**

5. a) Write an equation for the reaction of barium with dilute hydrochloric acid. Include state symbols. **[2 marks]**

   b) Describe what would be **seen** if a piece of barium were added to water. **[2 marks]**

   c) What colour would be seen if universal indicator solution were added to the aqueous product of the reaction in part **b)**? **[1 mark]**

DAY 5 — 60 Minutes

(orbitals?)
a shell can have different electron levels

Group 7: The Halogens

Halogens are non-metals with 7 electrons in their outer shell. The outer electrons are s^2, p^5.

They are only slightly soluble in water but readily dissolve in organic solvents, e.g. hexane.

	Standard state	Colour in water	Colour in organic solvent
Cl_2	green gas	green	green
Br_2	brown liquid	red/brown	brown
I_2	grey solid	yellow/brown	purple

Halogens form diatomic covalent molecules F_2, Cl_2, Br_2, I_2.

chlorine molecule

When they react they are typically reduced to form **halide** ions with a 1– charge.

This means they are oxidising agents, removing electrons from other species.

Changes Down the Group

F	Melting point increases ↓	Atomic radius increases ↓	Electronegativity decreases ↓	Oxidising ability decreases ↓
Cl				
Br				
I				
As				

Melting point and boiling point increase down the group. As the number of electrons increases so instantaneous dipole–induced dipole forces between molecules increase and more energy is needed to break molecules apart (see page 36).

Iodine forms solid crystals at room temperature. The crystals sublime (turn directly from a solid to a gas without being a liquid) on heating as the bonds holding the molecules together are weak.

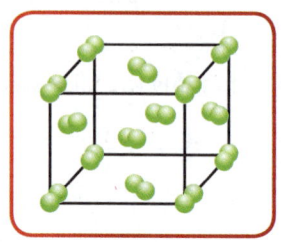

Reactivity decreases down the group. Each subsequent halogen has one additional shell of electrons. The outer shell of electrons gets further from the positive pull of the nucleus and has more shells of electrons shielding the nucleus, and is therefore less able to attract electrons. So, oxidising ability decreases down the group.

removing electrons

Predicting Properties

The property of a halogen can be predicted from the trends down the group. Astatine, At, is the halogen with the highest atomic mass. It is very rare and radioactive, but its properties can be predicted.

Colour	black
State	solid
Melting point	higher than iodine
Oxidising ability	lower than iodine

Compare the oxidising ability of halogens by mixing halogens with halide ions.

A more reactive halogen will displace a less reactive halogen from its compound in solution. The reactions are easy to see because of the different colours of the halogens. The reaction takes place in aqueous solution but the colour of the products can be made more obvious by adding an organic solvent layer. The halogen transfers from the aqueous layer to the organic layer because halogens are more soluble in organic solvents. *(than water)*

The ionic equations for the reactions:

Chlorine oxidises bromide ions.

$$Cl_2(aq) + 2Br^-(aq) \rightarrow Br_2(aq) + 2Cl^-(aq)$$

Colour changes from green to brown. Adding an organic solvent gives a brown organic layer on top of the water.

72

halides – the anionic form of halogens, do not have any unpaired electron (so gain one usually to have -1 charge)

Chlorine oxidises iodide ions.

$$Cl_2(aq) + 2I^-(aq) \longrightarrow I_2(aq) + 2Cl^-(aq)$$

Colour changes from green to brown. Adding an organic solvent gives a purple organic layer.

Bromine oxidises iodide ions.

$$Br_2(aq) + 2I^-(aq) \longrightarrow I_2(aq) + 2Br^-(aq)$$

Colour changes from red/brown to yellow/brown. Adding an organic solvent gives a purple organic layer.

Bromine and iodine are extracted from seawater by displacement with chlorine.

Reactions of Group 7 with Group 2

Metals are readily oxidised while the halogen is reduced to produce an ionic solid, e.g.

$$Ca(s) + Cl_2(g) \longrightarrow CaCl_2(s)$$

Oxidation number: 0   0   +2 −1

ionic bond

Disproportionation of chlorine: chlorine is oxidised and reduced in the same reaction.

Chlorine + water reacts to give hydrogen chloride and chloric(I) acid.

Oxidation State Cl $Cl_2 + H_2O \longrightarrow HCl + HClO$
       0   −1   +1

Chlorine is oxidised $0 (Cl_2) \longrightarrow +1 (HClO)$
and   reduced $0 (Cl_2) \longrightarrow -1 (Cl^-)$

Chlorine + cold dilute NaOH reacts to give sodium chloride, sodium chlorate(I) and water.

$$Cl_2 + 2NaOH \longrightarrow NaCl + NaClO + H_2O$$

This is also disproportionation of chlorine.

NaClO is the active ingredient in bleach.

Chlorine + hot concentrated NaOH reacts to give sodium chloride, sodium chlorate(V) and water.

$$3Cl_2 + 6NaOH \longrightarrow 5NaCl + NaClO_3 + 3H_2O$$
   0   −1   +5

Cl is +1 ??

Reactions of Halide Ions

Reducing ability of the halides increases down the group. Iodide ions donate electrons better than bromide ions, which donate better than chloride ions. This is reflected in their ability to reduce SO_4^{2-} ions from sulfuric acid.

Halides + Concentrated H_2SO_4
Chloride ions:

$$NaCl + H_2SO_4 \longrightarrow NaHSO_4 + HCl$$

Chloride is not a sufficiently good reducing agent to reduce sulfate. The reaction is **displacement not redox**. There are no changes in oxidation state. This reaction is used in the preparation of HCl.

Bromide ions:

$$2NaBr + 2H_2SO_4 \longrightarrow Na_2SO_4 + SO_2 + Br_2 + 2H_2O$$

Br^- is oxidised to Br_2; S is reduced from +6 to +4.

Iodide ions:

$$8NaI + 5H_2SO_4 \longrightarrow 4I_2 + H_2S + 4H_2O + 4Na_2SO_4$$

I^- is oxidised to I_2; S is reduced from +6 to −2.

Halides + silver nitrate give a precipitate of silver halide.

$$X^-(aq) + Ag^+(aq) \longrightarrow AgX(s)$$
$$NaCl(aq) + AgNO_3(aq) \longrightarrow AgCl(s) + NaNO_3(aq)$$

The reaction is **ionic precipitation not redox**.

This is used as a test for the presence of halide ions.

Cl^- gives white AgCl solid, which dissolves in dilute and concentrated ammonia solution.

Br^- gives cream AgBr solid, which dissolves only in concentrated ammonia solution.

I^- gives yellow AgI, which does not dissolve in ammonia solution.

Don't forget

…to describe the reason for halogen displacement reactions in terms of relative oxidising ability of the halogen rather than reactivity.

…to always show state symbols for ionic precipitation to demonstrate the production of a solid.

73

DAY 5   *anomalous = abnormal*

Hydrogen Halides

$$HF(g) \quad HCl(g) \quad HBr(g) \quad HI(g)$$

Properties

All halogen halides are gases at room temperature and pressure. Their boiling points increase down the group from HCl to HI. HF is anomalous because it has hydrogen bonds between the molecules whereas the other hydrogen halides have van der Waals and dipole–dipole forces.

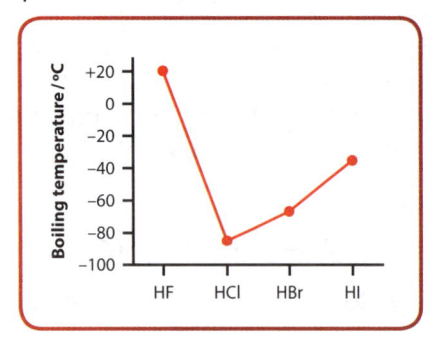

They are covalent molecules as gases but can dissolve in water to form ions (see below).

The strength of the hydrogen–halogen bond decreases down the group as the radius of the halogen gets bigger and the bond gets longer.

The polarity of the bond deceases down the group as the halogens become less electronegative.

Preparation of Hydrogen Halides

HCl can be prepared by reaction of NaCl with concentrated sulfuric acid (see previous page).

Other hydrogen halides are prepared by reaction of a halide with phosphoric acid.

$$NaBr(aq) + H_3PO_4(aq) \longrightarrow HBr(aq) + NaH_2PO_4(aq)$$

Reaction with Ammonia

All hydrogen halides react with ammonia to form an ammonium salt.

$$HX(g) + NH_3(g) \longrightarrow NH_4X(s)$$

Thermal Decomposition of Hydrogen Halides

Hydrogen halides can be broken back down to hydrogen and halogen by heating.

$$2HX(g) \longrightarrow H_2(g) + X_2(g)$$

Thermal stability decreases down the group as the bond enthalpy decreases. HF and HCl require high temperatures.

Acidity of Hydrogen Halides

All hydrogen halides are very water soluble and dissolve to produce acids. *split to form new things*

$$HX(g) \xrightarrow{H_2O} H^+(aq) + X^-(aq)$$

HCl, HBr and HI all fully dissociate and are strong acids. The strength of the acid decreases from HCl(aq) to HI(aq). HF only partially dissociates so is classified as a weak acid.

Acids and pH

Acids are $H^+(aq)$ donors and bases are $H^+(aq)$ acceptors, e.g.

$$HCl + NaOH \longrightarrow NaCl + H_2O$$

The HCl donates H^+ to the OH^- ion.

A strong acid fully dissociates in solution.

$$HCl \longrightarrow H^+ + Cl^-$$

A weak acid is in equilibrium with only a small percentage of the molecules dissociated.

$$CH_3COOH \rightleftharpoons CH_3COO^- + H^+$$

The pH of an acid shows the concentration of hydrogen ions in solution.

$$pH = -\log[H^+] \quad [H^+] = 10^{-pH}$$

pH of 0.1 mol dm^{-1} = $-\log 0.1$ = pH 1

What is the concentration of HCl with pH 3.0?

$$[H^+] = 10^{-pH} = 10^{-3.0} = 0.001 \text{ mol dm}^{-3}$$

QUICK TEST

1. What colour is iodine when dissolved in an organic solvent?
2. Explain why the boiling point of the halogens increases down the group.
3. Predict what would be seen if chlorine solution was mixed with potassium iodide solution.
4. Write an equation for the reaction of chlorine with water.
5. Write an equation for the reaction of sodium bromide with concentrated sulfuric acid.
6. Which halide ion gives a cream precipitate with silver nitrate?

enthalpy – sum of system's internal energy + product of P,V

PRACTICE QUESTIONS

1. The reaction that occurs when bromine solution is added to potassium iodide is [1 mark]

   A a redox reaction ☐   B a precipitation reaction ☐
   C an acid base reaction ☐   D There is no reaction. ☐

2. Which statement is true? [1 mark]

   A Bromine is a better oxidising agent than chlorine. ☐
   B Bromide is a better reducing agent than chlorine. ☐
   C Iodide is a better oxidising agent than chloride. ☐
   D Iodine is a better oxidising agent than chlorine. ☐

3. The reason for the decreasing thermal stability of hydrogen halides going down the group is [1 mark]

   A the increasing polarity of the covalent bond ☐
   B the decreasing radius of the halogen ☐
   C the increasing bond length ☐
   D the increasing bond enthalpy. ☐

4. HF(aq) is a weaker acid than HCl(aq). Which statement is correct? [1 mark]

   A The pH of 1 mol dm^{-1} HF will be lower than the pH of 1 mol dm^{-1} HCl. ☐
   B The concentration of hydrogen ions will be lower in 1 mol dm^{-1} HF than in 1 mol dm^{-1} HCl. ☐
   C HF is more highly dissociated than HCl. ☐
   d. A more concentrated solution of HF is a strong acid. ☐

5. When chlorine water is added to potassium iodide a colour change from green to brown is seen.
   a) Use an equation to explain the reaction. [2 marks]
   b) Chlorine is able to displace bromine from potassium bromide but iodine cannot. Explain why in terms of the atomic structure of the halogens. [3 marks]
   c) Iodide ions are able to reduce concentrated sulfuric acid to hydrogen sulfide but chloride ions cannot. Use chemical equations to show the difference between the reactions. Give the oxidation states of the halogens. [4 marks]

6. Bleach is prepared by reacting chlorine with sodium hydroxide solution. It can be used as a disinfectant because of its powerful oxidising ability, which kills microorganisms.
   a) Write an equation for the reaction of chlorine with cold sodium hydroxide. [2 marks]
   b) This reaction is known as a disproportionation reaction. Explain what is meant by a disproportionation reaction and explain why the reaction in part **a)** is classified in this way. [3 marks]
   c) Identify which of the products is responsible for its oxidising ability. [1 mark]

7. Describe the trend in boiling point of the hydrogen halides and explain why the boiling point of hydrogen fluoride is anomalous. [3 marks]

75

Uses of Group 2 and 7 Elements and Compounds

Uses of Calcium Compounds

Calcium carbonate, calcium oxide and calcium hydroxide are bases.

Increasing Soil pH

Many crops require a soil close to neutral if they are to give a high yield. Acidity in soil may be due to acid rain, natural breakdown of plant material producing hydrogen ions, or high crop yields that remove basic ions from the soil faster than they are replaced by the surrounding rocks.

Limestone is mainly $CaCO_3$. When added to soil the Ca^{2+} ions are slowly released and replace the H^+ which is bound to the soil particles. The released H^+ is neutralised by the carbonate ions.

$$2H^+ + CO_3^{2-} \rightleftharpoons CO_2 + H_2O$$

Calcium hydroxide is added to soil for the same purpose. It contains more Ca^{2+} per kg because of its lower molar mass. This reduces transport energy costs.

Neutralising Flue Gases

In power stations SO_2 can be removed by spraying a solution of $Ca(OH)_2$ through the waste gases.

$$2Ca(OH)_2 + 2SO_2 + O_2 \longrightarrow 2CaSO_4 + 2H_2O$$

The calcium sulfate produced (gypsum) can be used in the building industry to make products like plaster board.

In Medicines

$CaCO_3$ and $Mg(OH)_2$ are used as antacids for indigestion and heartburn. They neutralise excess hydrochloric acid produced by the stomach without producing a pH that is too high for living tissue.

$$CaCO_3 + 2HCl \longrightarrow CaCl_2 + H_2O + CO_2$$

$$Mg(OH)_2 + 2HCl \longrightarrow MgCl_2 + 2H_2O$$

Use of Barium Compounds as Barium Meal

Barium sulfate is insoluble and cannot be penetrated by X-rays. Patients are given a suspension of barium sulfate to swallow, which passes through the gastrointestinal tract making it opaque to X-rays. This gives a silhouette of the gut, highlighting any abnormal narrowing due to tumours or irregularities due to ulcers.

Barium salts are toxic but the insolubility of barium sulfate means that almost none is absorbed from the gut.

Chlorine as a Water Steriliser

Chlorine is a good oxidising agent that kills microorganisms in the water making it safe to drink. Opinion differs as to whether the benefits of using chlorine to sterilise water outweigh the risks.

Advantages of chlorine use:

Treating water with chlorine eliminates life-threatening waterborne diseases such as cholera and typhoid.

Chlorine treatment is cheaper than alternative methods such as ozone.

Disadvantages of chlorine use:

Use and transport of chlorine gas poses a potential health risk because of its toxicity and the danger of escaping gas. Gas spills cannot be contained – a spillage of gas is much more difficult to control than a spillage of a liquid or a solid.

Some chorine reacts with naturally occurring organic material to form chloroalkanes such as trichloromethane. These have been implicated in increased cancer rates and other health problems.

Fluoride in Drinking Water

1 mg of fluoride per dm^3 of drinking water has been shown to reduce tooth decay. In areas where the natural level of fluoride is lower than this, it can be

artificially added to drinking water. In the UK each local authority makes the decision about whether to fluoridate the water.

Advantages of adding fluoride to water:
Studies show that tooth decay has decreased in areas that have added fluoride to the water.

Disadvantages of adding fluoride to water:
Some studies suggest that tooth decay has also decreased in areas that do not have added fluoride. There have been no clinical trials on the use of fluoride, unlike other drugs.

Ingesting fluoride has been proposed as a factor in a variety of health problems. All authorities agree that high levels of fluoride lead to fluorosis. Fluorosis is a discolouration of teeth, usually flecks or lines in the tooth enamel, but causing pitting or brown stains in severe cases.

Those who oppose fluoridation of drinking water say that putting fluoride in drinking water is mass medication and denies people the choice of whether to take fluoride or not. This makes it unethical. True

Summary of Inorganic Tests

Flame Tests for Metal Ions

Group 1 and 2 metal ions give specific colours when heated in a flame. Electrons in the ions absorb heat energy and are excited from their ground state to higher energy levels. As the electrons drop back down to their ground state they emit light energy in the visible region. The wavelength of light emitted depends on the energy gap between the excited and ground states and is different for each element. Different wavelengths of light are seen as different colours.

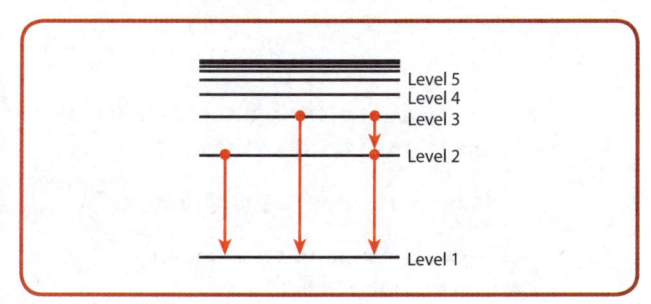

Excited electrons drop back to lower energy levels releasing light of specific wavelength.

Flame Colours

Ion	Li^+	Na^+	K^+	Rb^+
Flame colour	red	yellow	violet	red/violet

Ion	Ca^{2+}	Sr^{2+}	Ba^{2+}	*Cu^{2+}
Flame colour	brick red	crimson red	green	green blue

* Not Group 1 or 2 but useful to know.

Mg does not give a coloured flame.

How to carry out a flame test:

1. Dip a nichrome wire in concentrated hydrochloric acid and hold it in a hot Bunsen flame to clean. Repeat until no colour is produced from the wire.

2. Wet the wire again with the acid and dip into the solid sample.

3. Place the wire in the flame.

4. Note the flame colour produced.

Testing for Halide Ions

Halide ions form precipitates when mixed with silver ions. Different silver halide salts have different colours and solubilities in ammonia solution.

$$X^- (aq) + Ag^+(aq) \longrightarrow AgX(s)$$

Halide	Cl^-	Br^-	I^-
Colour of silver salt	white	cream	yellow
Solubility in ammonia solution	soluble	only soluble in concentrated ammonia solution	insoluble

DAY 5

How to carry out a test for halide ions:

1. Dissolve solid samples in water.
2. Acidify with nitric acid to remove other ions that may give a precipitate (carbonate ions).
3. Add a small amount of silver nitrate solution and look for a precipitate.
4. Add dilute ammonia solution to the precipitate and shake to dissolve.
5. If the precipitate remains, add concentrated ammonia solution and shake.

Note: This test does not identify F^- since it does not give a precipitate.

Testing for Cations with Hydroxide Ions

Most metals ions form insoluble hydroxides and some have characteristic colours.

$$M^{n+}(aq) + nOH^-(aq) \longrightarrow M(OH)_n(s)$$

Metal ion	Colour of precipitate	Formula
Cu^{2+}	blue	$Cu(OH)_2$
Fe^{2+}	green	$Fe(OH)_2$
Fe^{3+}	brown	$Fe(OH)_3$
Ca^{2+} Al^{3+} Zn^{2+} Mg^{2+}	white	

To distinguish between white precipitates:
- add excess sodium hydroxide solution: $Al(OH)_3$ dissolves
- add ammonia solution: $Zn(OH)_2$ dissolves
- calcium gives a flame colour, magnesium does not.

Testing for Ammonium Ions

Ammonium ions react with sodium hydroxide to give off ammonia gas.

$$NH_4^+(aq) + OH^-(aq) \longrightarrow NH_3(g) + H_2O(l)$$

Ammonia is an alkaline gas that turns damp red litmus paper blue.

Test for Acids and Carbonates

Acids release carbon dioxide from carbonates.

$$2H^+(aq) + CO_3^{2-}(aq) \longrightarrow CO_2(g) + H_2O(l)$$

Add sodium carbonate to the suspected acid. Look for effervescence. Bubble the gas released through limewater. A white precipitate indicates carbon dioxide.

$$CO_2(g) + Ca(OH)_2(aq) \longrightarrow CaCO_3(s)$$

Note: A pH probe or indicator would also show acidity.

Add any acid to a suspected carbonate or hydrogen carbonate.

To distinguish between carbonates and hydrogen carbonates heat a sample of the solid and check for release of carbon dioxide. Hydrogen carbonates decompose readily; carbonates are more thermally stable.

Test for Sulfate Ions

Barium and sulfate ions react to give a white precipitate.

$$Ba^{2+}(aq) + SO_4^{2-}(aq) \longrightarrow BaSO_4(s)$$

Test a suspected sulfate with acidified barium chloride solution. The solution is acidified to ensure that SO_3^{2-} ions do not give a precipitate.

QUICK TEST

1. Suggest why crushed limestone is sometimes added to soil.
2. Write an equation showing the reaction of calcium hydroxide with sulfur dioxide and oxygen.
3. List the properties of barium sulfate that make it suitable for use in barium meals.
4. What flame colour does barium produce?
5. Describe what happens when sodium hydroxide is added to $FeCl_2$.
6. Describe the test for chloride ions.

PRACTICE QUESTIONS

1. A student carried out a test for bromide ions by adding acidified silver nitrate solution to the suspected halide ion solution. The outcome was an off-white precipitate. The student could confirm the presence of the bromide ions by **[1 mark]**

    A adding excess sodium hydroxide and looking for a cream precipitate ☐

    B adding dilute ammonia solution and seeing if the precipitate dissolved ☐

    C checking solubility in concentrated and dilute ammonia solution ☐

    D leaving the precipitate exposed to light and seeing if it got darker. ☐

2. Some metal cations show characteristic colours when put into a flame. Which is the correct explanation for the colour? **[1 mark]**

    A Electrons are excited to higher energy levels and absorb light. ☐

    B The metal ions become oxidised. ☐

    C Electrons are given sufficient energy to escape from the pull of the nucleus. ☐

    D Excited electrons fall back to lower energy levels. ☐

3. Limestone has been used as a soil improver since before the 16th century. The limestone is crushed before adding to the soil and acts over a long period of time. Modern agriculture makes use of calcium hydroxide to take the same role. It is faster acting but requires more processing.

    a) How does soil become acidified? **[1 mark]**

    b) Write an equation for the reaction of limestone with H^+ ions. **[2 marks]**

    c) How can calcium hydroxide be produced from limestone? **[1 mark]**

    d) What advantage does calcium hydroxide have over limestone as a soil improver? **[1 mark]**

4. Chlorine is used in the water purification process.

    a) What is the function of chlorine in water purification? **[1 mark]**

    b) Why are there some concerns about the use of chlorine in water purification? **[2 marks]**

    c) What disadvantages would there be if chlorine were no longer used in water purification? **[2 marks]**

5. Students carried out a series of tests to identify an unknown ionic compound and obtained the results shown below. Each test was conducted on a fresh sample of the compound.

Test	Result
flame test	no flame colour
add sodium hydroxide to a solution of the compound	white precipitate
add excess sodium hydroxide	precipitate dissolves
add silver nitrate solution acidified with nitric acid	no change observed
add hydrochloric acid	no change observed

    a) Identify the cation present in the compound from the results of the tests and explain your choice. **[2 marks]**

    b) Eliminate some anions from the results and explain your answer. **[2 marks]**

    c) Suggest a further test that could be used to help identify the anion present. **[1 mark]**

DAY 5 60 Minutes

Chemistry and the Environment

Alternative Fuels

Carbon (in the form of coal, or in hydrocarbons) that has been buried underground for many millions of years is released as carbon dioxide into the atmosphere. The percentage of carbon dioxide in the atmosphere has increased over the last century. Carbon dioxide is a greenhouse gas that may contributes to climate change.

Fuels made from plant materials release less carbon dioxide to the atmosphere over their lifecycle because as the plants grow they absorb carbon dioxide by photosynthesis. The **net** carbon dioxide release is therefore less than that of fossil fuels.

Biodiesel

Biodiesel can be made from plant oils. The oils are modified by a process called transesterification, which converts triesters in the oil into methyl esters.

Advantages:
- The net carbon dioxide release is lower for biodiesel than for mineral diesel from crude oil.
- Existing diesel engines can use biodiesel without modifying the engine.
- The source of the fuel is renewable because more plants can be grown.
- Biodiesel is fully biodegradable so any spills are less damaging to the environment than conventional fossil fuels.
- Waste oils such as used chip shop oil can be used to make biodiesel.

Disadvantages:
- Biodiesels are not carbon neutral because energy is used in fertiliser manufacture, and in farming and processing the plant oil.
- Biodiesel is more expensive to produce than current fossil fuels.
- There is not enough land available to enable biodiesel to meet the world demand for diesel.
- Land used to grow plants for biodiesel is no longer available for food production.
- The rubber seals in the engine need to be replaced as biodiesel will make them perish. If the engine has used mineral diesel and then uses biodiesel the filters will become blocked up.

Bioethanol

Bioethanol is ethanol made from plant sources, usually by fermentation of sugars.

The starting material is plants high in sugar such as sugar cane. The cane is crushed and the sugar extracted with water. Yeast is used to convert the sugar into ethanol.

$$C_6H_{12}O_6 \longrightarrow 2CH_3CH_2OH + 2CO_2$$

Advantages:
- The net release of carbon dioxide is lower than for fossil fuels.
- The source is renewable because more plants can be grown.
- Bioethanol can be mixed with conventional fuels up to about 10% and used in petrol engines without modification.
- It improves the octane number of petrol and is less toxic than other petrol components used to improve octane number.

Disadvantages:
- The production is slow requiring warm, anaerobic conditions to allow yeast to grow and ferment the sugars.
- The fermentation product is a mixture and the ethanol must be removed by distillation.
- Energy is used in preparing the land and harvesting the crop as well as in the distillation process. Using ethanol as fuel is therefore not carbon neutral.
- Bioethanol is more expensive than petrol from crude oil.
- Bioethanol is less energy dense than petrol so requires approximately one third more fuel to travel the same distance.
- Using more than a 10% mixture with petrol requires modification of the car engine.
- The cutting down of trees to provide land for growing plants is damaging wildlife habitats, e.g. loss of rainforest in Brazil for sugar cane plantations.

Hydrogen

Hydrogen is an energy dense fuel that releases large amounts of energy when burned.

$$2H_2(g) + O_2(g) \longrightarrow 2H_2O(l) \quad \Delta H = -286 \text{ kJ mol}^{-1}$$

Hydrogen can be produced by electrolysis of water.

$$2H_2O(l) \longrightarrow 2H_2(g) + O_2(g)$$

Advantages:
- Burning hydrogen does not produce polluting gases.
- It does not release carbon dioxide.
- Hydrogen has a very high energy density by mass (142 kJ g^{-1}) compared to petrol (46.4 kJ g^{-1}).
- It does not use up a finite resource as hydrogen is generated from water, which is also the product of combustion.
- A network of pipes to transport gas already exists that could be used for transporting hydrogen.

Disadvantages:
- Electricity is required to produced hydrogen. Most electricity is currently produced by burning fossil fuels. Renewable energy would have to be used to give an advantage over direct combustion of fossil fuels.
- It is relatively expensive to electrolyse water by renewable energy.
- Current vehicles and fuelling stations cannot be used directly for hydrogen.
- Hydrogen has a very low energy density in volume terms because hydrogen is a gas. To give a suitable density for transport vehicles it must be stored under pressure. Pressure-resistant containers are heavy so make the transport vehicles heavy. There is a risk of explosion if pressurised containers leak.

Fuel Cells

Fuel cells produce electricity from the reaction between hydrogen and oxygen. They are a type of battery that do not run flat because they are continually supplied with the fuel required. They are very efficient at transferring the energy of the reaction to electricity compared to burning fossil fuels. The technology is well developed and has been used for decades in the space programme. Electric cars have been manufactured that use fuel cells as an energy supply but they are currently expensive.

Problems wih fuel cells are the same as those of using hydrogen as a fuel by combustion. They require a supply of hydrogen and this must be stored under pressure.

Problems with Polymers

Polymers made from crude oil are using up a finite resource.

Artificially produced polymers are not biodegradable and so persist for a long time in the environment. This has resulted in plastic waste taking up large amounts of land in landfill sites, as well as unsightly litter and damage to animals and habitats.

Chemists are part of the solution to the problem of what to do with polymer products that are no longer useful.

Recycling

Polymers can be collected and recycled. Thermosoftening polymers can be remelted and reshaped, but not all polymers will blend together. In order to give a useful product the polymers must be sorted into types. This makes recycling an expensive option. Recycled polymers from general waste are not pure and are of mixed colour. This means the products made are dark in colour and less strong than pure polymers.

Single polymer waste from manufacturers can be broken back down to monomers and these can be used to make new high quality polymers.

Cracking

Methods have been developed to break polymers down to smaller molecules that can be used as feedstock in the chemical industry. With current technology this is costly and inefficient and the capital costs of setting up a production process are high.

Incineration

Polymer waste can be incinerated to generate energy. Incineration releases carbon dioxide into the atmosphere and sometimes produces toxic gases that must be removed from the waste gases before they are released to the environment.

DAY 5

polymer - material consisting of large molecules with many repeating subunits

Biodegradable Polymers

Polymers have been developed that can be broken down by bacteria. Starch molecules are incorporated into the polymer product, which is broken down after use and causes the polymer to break into small pieces.

C=O bonds within polymers absorb infrared energy, which causes the polymers to degrade more rapidly in the light.

Biobased Polymers

These are polymers made from or by plants. They do not use up finite resources and release less net carbon to the atmosphere than oil-based polymers. Most current biobased polymers are based on starch. These are biodegradable.

PLA is polylactic acid, a biobased polymer that has similar properties to conventional polymers and is used in packaging and disposable tableware. The monomer for PLA is lactic acid. This can be made from cornstarch or other root starches. Although the polymer is made from biological materials, it does not compost as easily as starch. An industrial composter is required for fast decomposition. PLA can be broken back down to its monomers and remade.

Principles of Green Chemistry

All industrial processes should aim to make the minimum negative environmental impact. Considerations should include:

- getting maximum energy efficiency by use of catalysts and choosing the reaction pathway with the lowest energy requirement
- minimising use of chemicals by reducing the number of steps in a reaction and choosing routes with a high atom economy (see page 53)
- maximising use of renewable feedstocks and designing products that break down after use rather than persist in the environment
- reducing hazardous chemicals by designing products that have the desired function with minimum toxicity; avoiding the use of solvents and minimising use of chemicals that could cause explosions or fires.

QUICK TEST

1. Explain why biofuels release less net carbon dioxide than fossil fuels.
2. State a disadvantage of bioethanol.
3. State an advantage of hydrogen as a fuel.
4. Explain why recycling polymers is expensive.
5. What problem arises with polymer incineration?
6. What is the feedstock for most biobased polymers?

PRACTICE QUESTIONS

1. Which of the following statements is not an advantage of biodiesel? **[1 mark]**

   A   It releases less carbon dioxide than fossil fuels.

   B   It has a higher energy density than fossil fuels.

   C   It is renewable.

   D   It is biodegradable.

2. Bioethanol has environmental advantages as well as disadvantages. Which of the following are true? **[1 mark]**

A	Bioethanol production is carbon neutral.	Production of bioethanol may damage ecosystems.
B	Bioethanol can be mixed with petrol.	Bioethanol releases carbon dioxide.
C	There is no need to adapt cars to use bioethanol.	Bioethanol is not renewable.
D	Bioethanol has a high energy density.	Bioethanol releases high levels of particulates.

3. In 1970 John Bockris proposed a system of delivery energy, which he called the hydrogen economy. This would be a replacement for the current hydrocarbon economy. At the moment the majority of hydrogen is produced from fossil fuels.

   a) What advantages would there be in exchanging hydrocarbons for hydrogen? **[3 marks]**

   b) Write an equation for the combustion of hydrogen and use it to explain why hydrogen can be considered a renewable fuel. **[2 marks]**

   c) Explain why the production of hydrogen from fossil fuels would reduce the advantages of a hydrogen economy. **[2 marks]**

   Another system for producing hydrogen is the electrolysis of water.

   Some people consider that changing to a hydrogen economy and using electrolysis to produce the hydrogen would significantly reduce the amount of carbon dioxide being released to the atmosphere.

   d) Explain why even a total hydrogen economy would not result in no carbon emissions. **[2 marks]**

   e) Hydrogen is already used as a rocket fuel. What difficulties have to be overcome in order to use it for cars? **[3 marks]**

4. a) Give one problem that has resulted from the use of synthetic polymers. **[1 mark]**

   b) Describe one solution to the problem of polymer waste and explain one drawback to this solution. **[2 marks]**

   c) Explain what is meant by a biobased polymer and describe two advantages of using such polymers. **[2 marks]**

DAY 6 / 60 Minutes

Organic Chemistry

Organic chemistry is the study of molecules that are made mainly of carbon and hydrogen. The **carbon skeleton** of organic molecules is not very reactive.

Functional groups attached to the skeleton determine the reactions of the molecule.

The reactions of a functional group remain similar whatever the carbon skeleton.

Homologous series are families of organic molecules with the same functional group. Each subsequent member has one additional CH_2 in the carbon chain. They show graduated changes in physical properties, similar chemical properties and the same general formula.

Naming Organic Molecules

Follow the C—C bonds to find the longest continuous carbon chain (regardless of orientation).

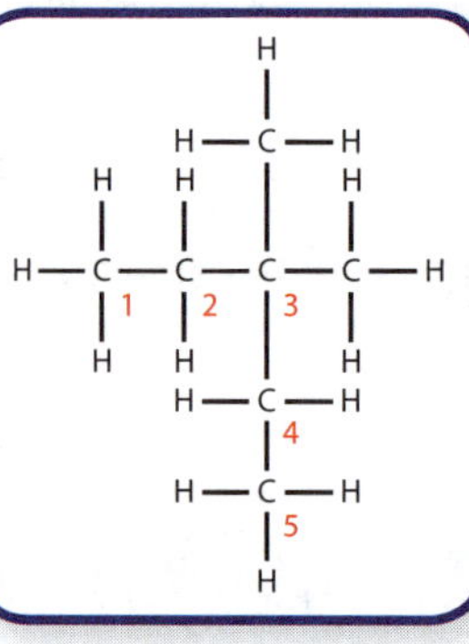

This gives the stem of the name.

Number of carbons	Name	Number of carbons	Name
1	meth	6	hex
2	eth	7	hept
3	prop	8	oct
4	but	9	non
5	pent	10	dec

The suffix (ending) of the name comes from the parent functional group.

Where there are no functional groups the molecules are alkanes. Alkanes have the suffix **-ane**, e.g. pent**ane**.

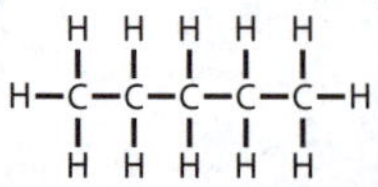

Prefixes are used to name and number any additional carbon chains attached to the longest chain (side groups). The number of carbons in each side group is shown using the system shown in the table above but with the ending **-yl**. Each side group is numbered, even if two side groups are attached to the same carbon. The carbon chain is numbered so as to give the lowest number to each side group.

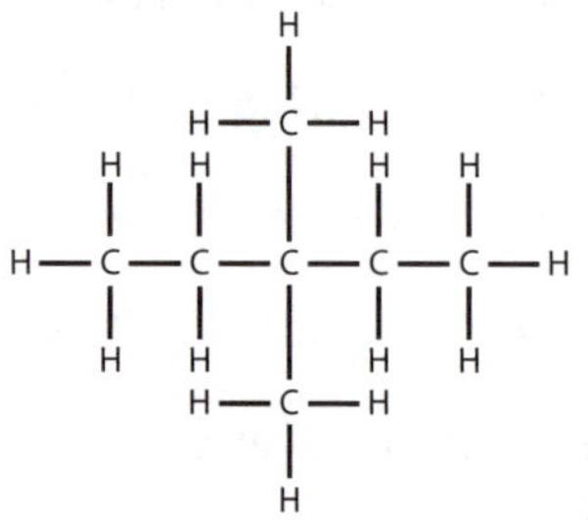

3,3-dimethylpentane

Separate numbers by a comma and put a hyphen between numbers and letters. Indicate more than one of the same group by using **di** (2), **tri** (3), **tetra** (4).

If there is more than one type of side group list them in alphabetical order.

Example
3-ethyl-2-methylhexane **not** 2-methyl-3-ethylhexane.

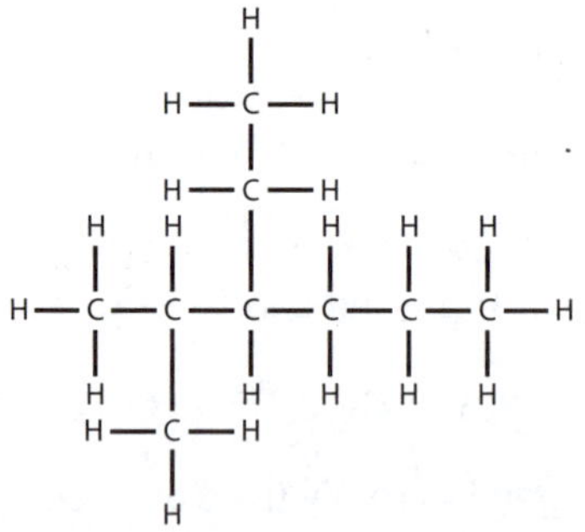

Naming Molecules with Functional Groups

Name the stem as alkanes but add a suffix or prefix (see table) and number the carbon to which the group is attached.

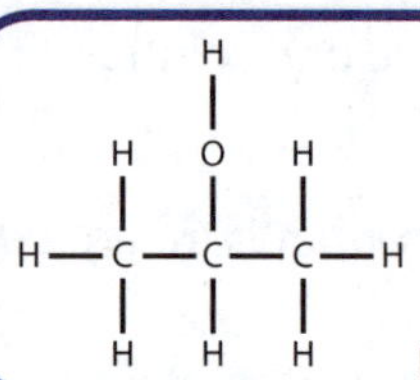

propan-2-ol

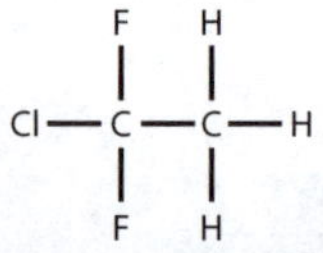

1-chloro-1,1-difluoroethane

The carbon chain is always numbered so as to give the lowest possible number to the functional group or side group.

Example
2-bromopentane **not** 4-bromopentane

Functional group	Type of compound	Suffix or prefix
C=C	alkene	-ene
–C–O–H	alcohol	-ol
–C–X	haloalkane	halo-
–C(=O)–H	aldehyde	-al
–C–C(=O)–C–	ketone	-one
–C(=O)–O–H	carboxylic acid	-oic acid
–C(=O)–O–C–	ester	-oate

Cyclic compounds use the prefix **cyclo-**.

Example
Cyclopentanol

Aromatic compounds contain a benzene ring. Hydrocarbons with benzene rings are called **arenes**.

Example
Phenylethane

A benzene ring is considered a functional group and takes the prefix **phenyl-**.

Aliphatic compounds do not contain a benzene ring.

Representing Organic Molecules

Displayed formulae show every bond and atom individually.

Example
2-methylpropan-1-ol

Structural formulae show the order of each group of atoms and functional group but not all bonds.

$$(CH_3)_2CHCH_2OH$$

Skeletal formulae show carbon–carbon bonds and functional groups only.

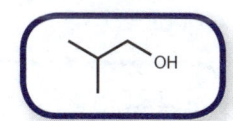

Molecular formulae give only number of each type of atom with no indication of structure.

$$C_4H_{10}O$$

Don't forget

…to give the molecular or skeletal formula if that is what is asked for – don't just write the structural formula.

…to draw the bond between O and H for displayed formula of alcohols (O—H not OH).

DAY 6

Organic Isomers

Structural isomers all have the same molecular formula but a different arrangement of atoms.

Chain isomers have different lengths of carbon chain and different lengths of side chains.

Example
3,3-dimethylpentane   2-methylhexane

Positional isomers have the same functional group but are in a different position along the chain.

Example
1-bromopropane   2-bromopropane

Functional group isomers have different functional groups.

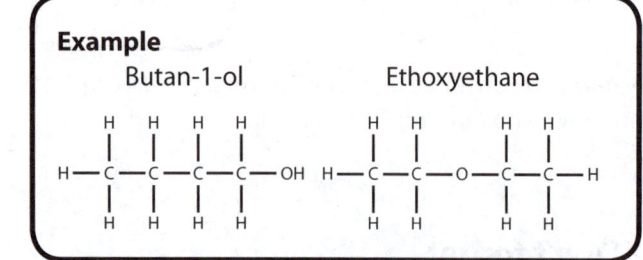

Example
Butan-1-ol   Ethoxyethane

Stereoisomers have the same structural formula but a different arrangement of atoms in space.

Example
But-2-ene

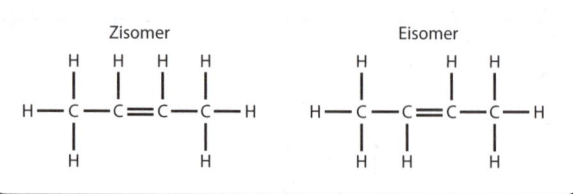

Zisomer   Eisomer

For details see Alkenes on page 96.

Don't forget

…to give all numbers for two or more halogens on the same carbon, e.g. 1,1-dichloro not 1-dichloro.

…to check you have identified the longest carbon chain – it might not lie horizontally.

…to count all hydrogen atoms when converting from skeletal to molecular formulae.

…to count the carbon chain using the lowest number carbon for each functional group or side chain – this might not be from the left.

QUICK TEST

1. Draw the displayed formula of 1,2-dimethylcyclopentan-1-ol.
2. Name this molecule.
3. Draw the skeletal formula of 1-bromo-2-methylpropane.
4. Write the molecular formula for this skeletal formula.
5. Draw the displayed formula for $CH_3CH(CH_3)CH_2CO_2H$.
6. Name a chain isomer of butane.
7. Name a positional isomer of 1,2-dichloroethane.

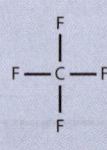

PRACTICE QUESTIONS

1. Which of the following is not an isomer of 2-methylpentane? **[1 mark]**

   A   3-methylpentane

   B   Cyclohexane

   C   2,3-dimethylbutane

   D   Hexane

2. The skeletal formula of limonene is

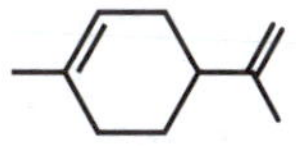

   The molecular formula of limonene is **[1 mark]**

   A C_9H_{14}

   B C_9H_{18}

   C $C_{10}H_{20}$

   D $C_{10}H_{16}$

3. CFCs, chlorofluorocarbons, have been banned from use in Europe because of their effect on the environment.

   a) To which class of compounds do CFCs belong? **[1 mark]**

   One CFC that was previously used is 1,2-dichloro-1,1,2,2-tetrafluoroethane.

   b) Draw the displayed formula of this molecule. **[1 mark]**

   c) Draw and name a positional isomer of 1,2-dichloro-1,1,2,2-tetrafluoroethane. **[2 marks]**

   HFCs are now being used as a less damaging alternative to CFCs. One common HFC has the formula CHF_3 and is usually known as fluoroform.

   d) What is the systematic name for fluoroform? **[1 mark]**

4. Fumaric acid is a compound commonly found in fruit.

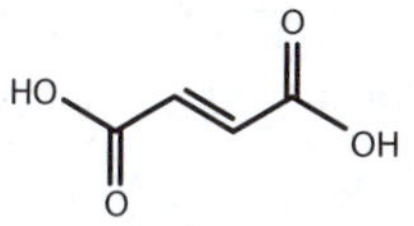

   a) Name all the functional groups seen in fumaric acid. **[2 marks]**

   The systematic name of a related compound, which is used as a flavouring, is pentenoic acid.

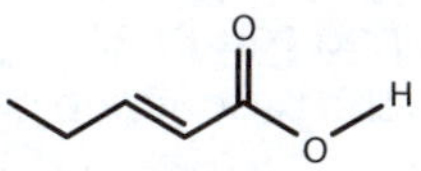

   b) Suggest the systematic name for fumaric acid. **[2 marks]**

   c) Show that fumaric acid and pentenoic acid are not isomers. **[3 marks]**

DAY 6 60 Minutes

Alkanes

Alkanes have the general formula C_nH_{2n+2}

Cycloalkanes have the general formula C_nH_{2n} $n \geq 3$ probs.

They are hydrocarbons – made from carbon and hydrogen **only**.

Alkanes are saturated; they have no double bonds. The C—H and C—C bonds are σ bonds. There is free rotation around the σ bonds.

Their boiling point increases with increasing chain length. This is due to increasing numbers of electrons and increased surface of contact between molecules and so increasing van der Waals forces.

	butane C_4H_{10}	pentane C_5H_{12}	hexane C_6H_{14}
b.pt °C	−1	36	69

Straight chain alkanes have higher boiling points than branched chain alkanes of the same M_r due to increased surface of contact.

2,2-dimethylpropane M_r 72 b.pt 10°C

pentane M_r 72 b.pt 36°C

Sources of Alkanes

Crude oil is a mixture of hydrocarbons.

Fractional distillation separates crude oil into useful fractions as each fraction has a different boiling point. The crude oil is vaporised and rises up a fractionating tower. The vapour cools as it rises and molecules condense as they reach their boiling temperature. Molecules with similar boiling points are tapped off together as fractions with different uses.

After fractionation the fractions may be processed in different ways to make them more useful.

Cracking is decomposition of long chain alkanes into shorter chain alkanes and alkenes. This converts the less valuable longer chain fractions into more useful shorter chain molecules, e.g.

$$C_{10}H_{22} \longrightarrow C_8H_{18} + C_2H_4$$

Thermal cracking uses high temperature and pressure, and yields a high percentage of alkenes, used for making polymers.

Catalytic cracking uses a zeolite catalyst and lower pressures. It yields many shorter and branched chain alkanes, which are useful in petrol. A zeolite is a giant network of silicon, oxygen and aluminium with an open structure.

Reforming is converting straight chain molecules into branched chain, cyclic, and aromatic compounds.

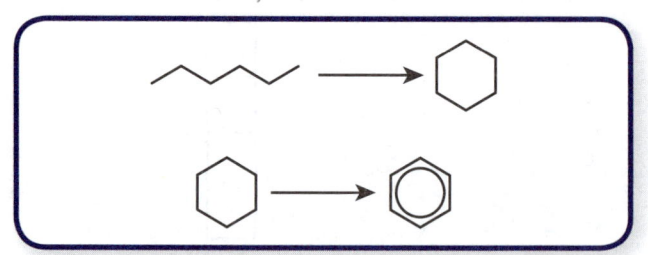

This gives petrol a higher octane number and reduces knocking in petrol engines. Octane number describes the tendency of petrol to autoignite.

Alkanes as Fuels

Alkanes from crude oil are known as fossil fuels because the original source of the carbon was from living organisms that became buried and fossilised millions of years ago. They are a finite resource since they will not be replaceable. Burning fossil fuels releases new carbon dioxide into the atmosphere from carbon that was previously underground.

Complete combustion is burning in plentiful oxygen. When hydrocarbons completely combust only carbon dioxide and water are made, e.g.

$$C_8H_{18} + 12.5O_2 \longrightarrow 8CO_2 + 9H_2O \quad \Delta H^\ominus = -5460 \text{ kJ mol}^{-1}$$

Alkanes are very energy dense and release large quantities of energy on combustion. They are the major source of energy for developed countries both by direct combustion and in the production of electricity in power stations.

Incomplete combustion, due to insufficient oxygen, produces a mixture of water, carbon dioxide, carbon monoxide and carbon. It releases less energy. Many different ratios of products are possible, e.g.

$$C_8H_{18}(g) + 9O_2(g) \longrightarrow 3CO_2(g) + 3CO(g) + 2C(s) + 9H_2O$$

Pollution from Burning Fossil Fuels

Polluting emissions when fossil fuels are burned include CO, NO_x, SO_2, C and unburned hydrocarbons.

CO is a toxic gas that prevents oxygen being transported by red blood cells.

SO_2 and NO_x are acidic gases that cause air pollution and acid rain. Both are associated with respiratory problems. NO is recognised as an asthma trigger.

The high temperatures and pressures in internal combustion engines and power stations cause the oxidation of $N_2(g)$ from the air to nitrogen oxide NO, as enough energy is provided to break the triple covalent bond in a nitrogen molecule.

$$N_2(g) + O_2(g) \longrightarrow 2NO(g)$$

Naturally occurring impurities of sulfur in crude oil cause SO_2 to be made when the fuel is burned.

$$S + O_2 \longrightarrow SO_2$$

Solid particulates of carbon can damage lung tissue and cause increased reflection of sunlight. This causes global dimming, which is thought to affect rainfall patterns.

Unburned alkanes and nitrogen oxides react together in the presence of sunlight to produce a mixture of secondary pollutants known as **photochemical smog**. One secondary pollutant is ozone, which causes damage to lungs and eyes.

Reducing Air Pollution

A catalytic converter allows NO and CO in waste gases from cars to react together to form gases naturally found in air, which are less harmful.

$$2CO + 2NO \longrightarrow N_2 + 2CO_2$$

It also catalyses oxidation of unburned hydrocarbons to CO_2 and H_2O.

The catalyst is mainly platinum with palladium and rhodium coated onto an inert honeycomb ceramic structure to provide a large surface area for the reaction.

The levels of SO_2 in the atmosphere have reduced in the last few decades since desulfurisation of diesel fuels was introduced.

SO_2 and NO_x can be removed from waste gases in power stations by reaction with CaO or $CaCO_3$ using a piece of equipment on the chimneys called a gas scrubber.

Reactions of Alkanes

Alkanes are relatively unreactive due to the high bond enthalpy and lack of polarity of C—H and C—C bonds.

Alkanes to Haloalkanes

React the alkane with a halogen in the presence of UV light, e.g.

$$CH_4 + Br_2 \longrightarrow CH_3Br + HBr$$

The mechanism is **(free) radical substitution** and takes place via a number of steps beginning with the production of a radical.

Initiation $Br_2 \longrightarrow 2Br\bullet$
Propagation $CH_4 + Br\bullet \longrightarrow CH_3\bullet + HBr$
                     $CH_3\bullet + Br_2 \longrightarrow CH_3Br + Br\bullet$
Termination $CH_3\bullet + Br\bullet \longrightarrow CH_3Br$

> **Radical:** A very reactive species with an unpaired electron.

Initiation: Homolytic bond fission arises when halogens are exposed to UV light.

$$Br\!-\!Br \longrightarrow Br\bullet \; Br\bullet$$

> **Homolytic fission:** Each atom receives one of the shared pairs of electrons to produce two radicals.

DAY 6

Propagation first step:

The radical reacts with the alkane.

$$Br\cdot + H-CH_3 \longrightarrow H-Br + \cdot CH_3$$

The alkane loses H and becomes an alkyl radical.

Propagation second step:

The alkyl radical reacts with a non-radical…

$$Br-Br + \cdot CH_3 \longrightarrow Br-CH_3 + Br\cdot$$

…producing a new radical.

Many different propagation steps are possible, e.g.

$$Br\cdot + H-CH_2Br \longrightarrow H-Br + \cdot CH_2Br$$

So, many different products are possible.

Termination step

Two radicals join together.

$$Rad\cdot + \cdot Rad \longrightarrow Rad-Rad$$

Many different terminations are possible.

Examples

$$H_3C\cdot + Cl\cdot \longrightarrow H_3C-Cl$$

$$H_3C\cdot + \cdot CH_3 \longrightarrow H_3C-CH_3$$

The many possibilities at the propagation and termination steps mean that this reaction always results in a mixture of products. The yield is never 100%.

QUICK TEST

1. Write the equation for the complete combustion of hexane.
2. List the conditions required for thermal cracking.
3. Explain why oil fractions are reformed.
4. Define the term free radical.
5. Explain why the radical substitution of alkanes only occurs in the presence of UV light.
6. Explain why the yield of a free radical substitution reaction is always less than 100%.

PRACTICE QUESTIONS

1. Which of the following are not the possible products of cracking of decane? [1 mark]

   A $2C_2H_4 + C_6H_{12}$ ☐ B $C_2H_4 + C_8H_{18}$ ☐

   C $C_6H_{14} + C_4H_8$ ☐ D $C_5H_{10} + C_5H_{12}$ ☐

2. Which of the following pollutants from car exhaust fumes are causes of acid rain? [1 mark]

   A NO_2 only ☐ B CO_2 ☐

   C CO ☐ D SO_2 and NO ☐

3. The alkanes present in crude oil are separated into useful fractions by fractional distillation.

   a) Why do the fractions separate from each other? [1 mark]

   b) If the following mixture of molecules was separated using fractional distillation, which would you expect to be removed at the lowest temperature? Explain your answer. [3 marks]

   A   B   C   D

   c) Name molecule **C**. [1 mark]

   d) Name the process that could be used to convert molecule **A** into **D**. [1 mark]

   e) Molecule **A** reacts with Br_2 in the presence of UV light. Write an equation for this reaction using molecular formulae. [2 marks]

   f) The reaction occurs in four stages. Give the name of the first step of the reaction mechanism. [1 mark]

   g) Explain why this reaction only occurs in the presence of UV light. [2 marks]

   h) What name is given to this reaction mechanism? [1 mark]

4. Chlorine reacts with methane to produce chloromethane. The first step is $Cl_2 \longrightarrow 2Cl\bullet$

   a) Show two propagation steps for this reaction. [2 marks]

   b) One side product of the reaction is ethane. Use a reaction equation and curly arrows to explain how ethane might form. [3 marks]

5. The majority of domestic vehicles are powered by an internal combustion engine using petrol. Petrol is a mixture of alkanes with a typical carbon chain length of 8. After combustion of the fuel, the waste gases leave the car via the exhaust pipe. These vehicles must comply with strict emissions tests if they are to be considered roadworthy. A catalytic convertor in the exhaust system can help to reduce the emission of polluting gases such as NO and CO.

   a) Write an equation showing the complete combustion of octane. [2 marks]

   b) Why are NO and CO considered to be polluting gases? [2 marks]

   c) What is the catalyst in a catalytic convertor? [1 mark]

   d) Use an equation to show how a catalytic convertor reduces the emission of polluting gases. [2 marks]

Haloalkanes (Halogenoalkanes)

Haloalkanes contain at least one of the following bonds:

Fluoroalkanes   C—F   Chloroalkanes   C—Cl

Bromoalkanes   C—Br   Iodoalkanes   C—I

Naming Haloalkanes

Haloalkanes follow the naming rules for alkanes but have a prefix of fluoro-, chloro-, bromo-, iodo-. The number of the carbon atom to which the halogen is attached is shown after the halogen name, e.g.

1-chloro-3,3-diiodobutane

Primary haloalkane: halogen is attached to a carbon that is attached to only one other carbon atom, i.e. CH_2, e.g. CH_3Br.

Secondary haloalkane: halogen is attached to a carbon that is attached to two other carbon atoms, i.e. CH, e.g. $CH_3CH(CH_3)Br$.

Tertiary haloalkane: carbon to which halogen is attached is not bonded to H, e.g. $CH_3C(CH_3)_2Br$.

Haloalkanes do not dissolve well in water so reactions are often completed in a mixture of water and ethanol.

Test for Haloalkanes

Warm the test solution with aqueous sodium hydroxide, neutralise with nitric acid and add silver nitrate. A precipitate indicates a haloalkane. White = chloroalkane; cream = bromoalkane; yellow = iodoalkane. Fluoroalkanes cannot be tested using this method as silver fluoride is soluble.

The Carbon–Halogen Bond

The C—halogen bond is polarised because of the differences in electronegativity between carbon and the halogen: $^{\delta+}C—X^{\delta-}$

The C—halogen bond gets less polar down the group as the electronegativity of the halogen decreases.

Atomic radius of the halogens increases down the group: F < Cl < Br < I. This means that the C—halogen bond length increases down the group.

The bond polarity is less significant than the bond length in deciding the bond enthalpy. Bond enthalpy decreases down the group as bond length increases.

Evidence for Decreasing Bond Enthalpy Down the Group

Heating a haloalkane in aqueous silver nitrate causes a halide ion to be released and a precipitate of silver halide to form. Under the same conditions, iodoalkanes form precipitates faster than bromoalkanes, which are faster than chloroalkanes.

In this reaction a tertiary haloalkane reacts faster than a secondary haloalkane, which reacts faster than a primary haloalkane.

Reactions of Haloalkanes

Nucleophile: A species that can donate a lone pair of electrons to form a covalent bond to carbon.

Examples: OH^-, NH_3, CN^-, H_2O

The δ+ carbon of haloalkanes attracts nucleophiles.

The polar C—halogen bond breaks by heterolytic fission to give a halide ion.

Heterolytic bond fission: Both electrons from the shared pair of a covalent bond go to one atom to produce ions.

The nucleophile and halogen swap places so the nucleophile is bonded to the carbon. This is known as a **nucleophilic substitution** reaction.

Nucleophilic Substitution Reactions of Haloalkanes

Haloalkane to alcohol: Reflux with an aqueous base.

$CH_3CH_2Br(l) + KOH(aq) \longrightarrow CH_3CH_2OH(aq) + KBr(aq)$

In aqueous solution OH$^-$ behaves as a nucleophile – donates a pair of electrons to form a covalent bond to carbon.

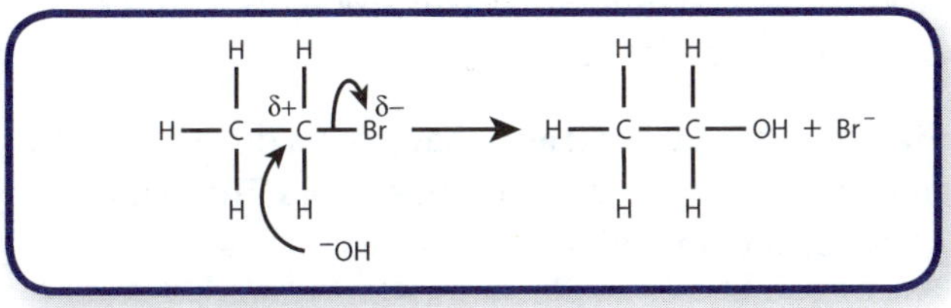

Water can also act as a nucleophile to produce an alcohol but the reaction is much slower.

Haloalkane to amine: Heat in a sealed tube with excess ammonia.

$CH_3CH_2Br + 2NH_3 \longrightarrow CH_3CH_2NH_2 + NH_4Br$

A sealed tube prevents gaseous ammonia from escaping. Excess ammonia prevents multiple substitutions on each ammonia.

Step 1

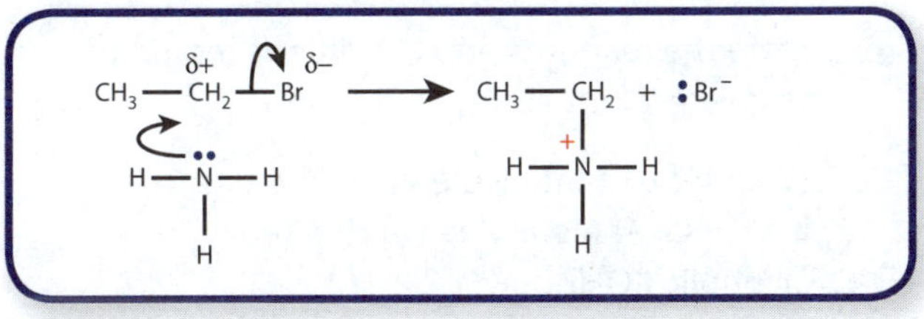

Step 2

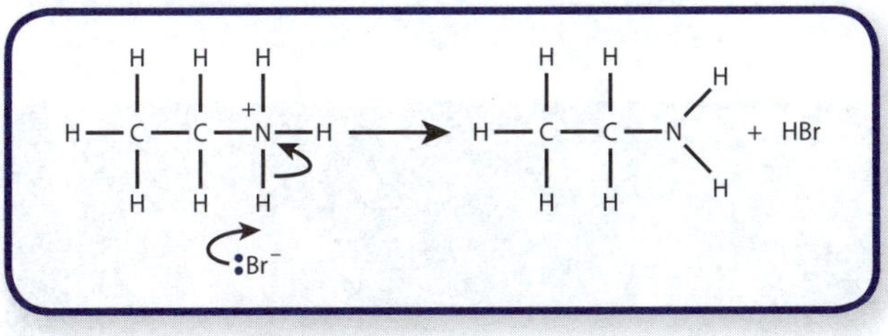

$HBr + NH_3 \longrightarrow NH_4Br$

Haloalkane to nitrile: Reflux with ethanolic KCN.

$CH_3CH_2Br + KCN \longrightarrow CH_3CH_2CN + KBr$

Ethanol is used as a solvent because water would result in production of an alcohol.

The reaction increases the carbon chain length by 1. bromo**eth**ane ⟶ **propane**nitrile

Elimination reactions

Haloalkane to alkene: Heat with an ethanolic strong base.

$CH_3CH_2Br + KOH \longrightarrow CH_2CH_2 + H_2O + KBr$

In alcoholic solution hydroxide behaves as a base – removes a proton (H$^+$).

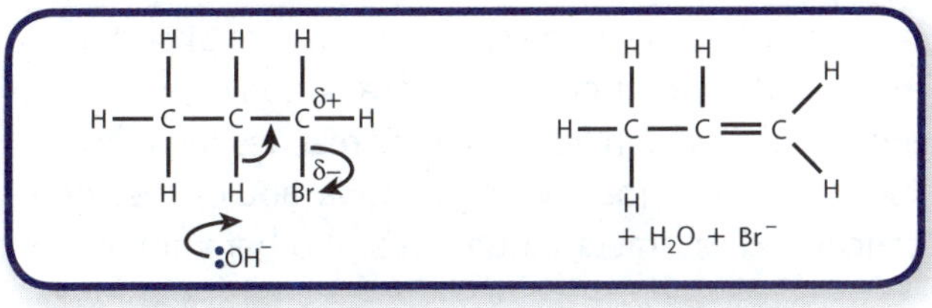

For secondary haloalkanes three possible isomers can form depending on which H is removed.

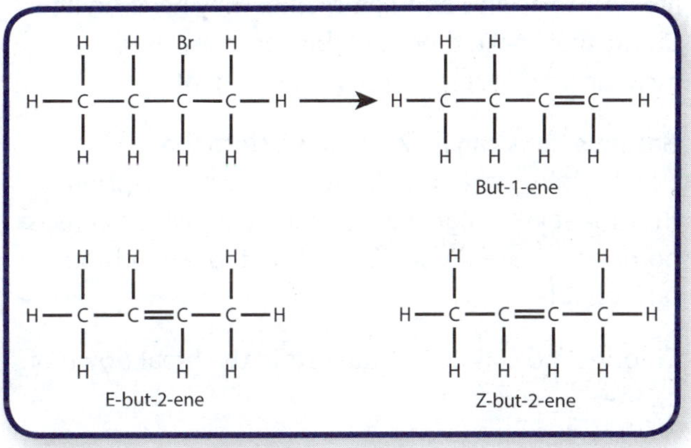

But-1-ene

E-but-2-ene   Z-but-2-ene

Competing Reactions

The substitution to alcohol and elimination reactions both occur in alkaline conditions. A mixture of products is likely.

High temperatures, high concentration of alkali, an ethanolic solvent and a tertiary haloalkane favour the elimination reaction.

Lower temperatures, dilute aqueous alkali and a primary haloalkane favour the substitution reaction.

93

DAY 6

Radical – molecule containing at least one unpaired electron

Haloalkanes and the Ozone Layer

Ozone Formation

In the stratosphere high level UV radiation from the sun causes homolytic fission of some oxygen molecules to form radicals.

$$O_2 \rightarrow 2O\bullet \quad \Delta H^\ominus = +498 \text{ kJ mol}^{-1}$$

The radicals react with more oxygen to form ozone.

$$O\bullet + O_2 \rightleftharpoons O_3 \quad \Delta H = -106 \text{ kJ mol}^{-1}$$

The ozone breaks back down when it absorbs UV so an equilibrium is established. This means that there is a layer of ozone in the stratosphere that absorbs much of the high frequency UV radiation coming from the sun. If this high frequency UV reaches the planet surface it can damage living material and cause skin cancer and cataracts in humans and other animals.

Problems with CFCs – Chlorofluorocarbons

Chlorofluoralkanes are very stable molecules that are non-toxic and non-flammable. They have boiling points that make them suitable for many uses including refrigerants and blowing agents.

Gaseous CFCs can make their way into the stratosphere unchanged. Once in the stratosphere they meet with high level UV radiation, which causes homolytic fission of the C—Cl bond to form chlorine radicals, Cl•.

Chlorine radicals act as catalysts in the breakdown of ozone.

$$O_3 + Cl\bullet \longrightarrow ClO\bullet + O_2$$

$$ClO\bullet + O_3 \longrightarrow 2O_2 + Cl\bullet$$

Overall reaction: $2O_3 \longrightarrow 3O_2$

The rate of breakdown of ozone with Cl• is 15 000 times faster than without Cl•. The chlorine is not used up in the reaction so one radical can catalyse the breakdown of hundreds of thousands of ozone molecules before reacting with something that removes it.

The presence of chlorine radicals has disturbed the natural equilibrium for the breakdown and formation of ozone and decreased the amount of ozone in the stratosphere. This has been linked with an increase in skin cancers.

CFCs have now been banned under the Montreal Protocol of 1987 but those that have already been released will continue to cause problems with the ozone layer for decades.

Hydrofluorocarbons (HFCs) have been developed that are possible to use in the place of CFCs. These do not cause ozone depletion.

QUICK TEST

1. Which haloalkane would have the highest boiling point, 1-bromobutane or 1-fluorobutane?
2. Describe how you would test for the presence of a suspected chloroalkane.
3. Draw the displayed structure of 2-chloro-2-methylpropane and state whether it as a primary, secondary or tertiary haloalkane.
4. State the reagents and conditions required to convert iodoethane to ethanol.
5. What type of molecule would be produced if a tertiary haloalkane was heated with alcoholic KOH?
6. What is the role of CFCs in the breakdown of ozone?

PRACTICE QUESTIONS

1. The following haloalkanes were each heated with silver nitrate solution.

    i. 2-iodopentane
    ii. 3-bromopentane
    iii. 1-chloropentane
    iv. 3-chloro-3-methylpentane

    The time order in which a precipitate formed would be [1 mark]

    A i then ii then iii then iv ☐
    B iv then iii then ii then i ☐
    C i then iv then ii then iii ☐
    D i then ii then iv then iii. ☐

2. An unknown haloalkane was warmed with sodium hydroxide and the resulting solution was acidified and then mixed with silver nitrate.

    Which statement is false? [1 mark]

    A Hydrochloric acid would be unsuitable to use for acidification. ☐
    B A white precipitate indicates the presence of a bromoalkane. ☐
    C If the haloalkane is a bromoalkane the precipitate will not dissolve in ammonia solution. ☐
    D The reaction of the haloalkane is a hydrolysis. ☐

3. a) Write a balanced equation for the reaction between iodopropane and ammonia. [2 marks]
   b) State the name of the homologous series to which the product belongs. [1 mark]
   c) State the reaction conditions required for this reaction. [2 marks]
   d) Name the reaction mechanism. [1 mark]

4. 2-bromohexane was heated with sodium hydroxide dissolved in ethanol to give hex-2-ene.
   a) Draw the displayed formula of the reactants and products and use curly arrows to show how the reaction proceeds. [4 marks]
   b) Explain why it is likely that a mixture of unsaturated products will be obtained. [1 mark]
   c) Explain why the products may also contain a saturated compound. [2 marks]

5. A student wanted to make pentanenitrile starting with a haloalkane. Starting materials available were bromoalkanes and fluoroalkanes.
   a) Which type of haloalkane should the student choose, bromo or fluoro, and why? [3 marks]
   b) State the reagents and conditions that would give the required product. [3 marks]
   c) Define the term nucleophile and give the formula of the nucleophile in this reaction. [3 marks]

6. In 1987 CFCs were banned under the Montreal Protocol.
   a) What are CFCs? [1 mark]
   b) Write an equation to show the formation and breakdown of ozone in the stratosphere. [1 mark]
   c) Use appropriate equations to describe the link between CFCs and stratospheric ozone. [4 marks]

DAY 6 — 60 Minutes

Alkenes

σ bond – strongest covalent bond, head-on overlap btwn orbitals
π bonds – lateral overlap btwn orbitals

Alkenes contain one or more C=C double bond.

Naming

Alkenes are named in the same way as alkanes but with the suffix -ene. The position of double bonds is indicated by the number of the first carbon atom of the double bond, using the lowest number possible. More than one double bond is shown by use of di, tri, tetra etc. before the –ene, e.g.

2-methyl-2,4-heptadiene
not 6-methyl-4,6-heptadiene

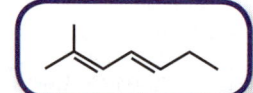

Bonding

The bond length of the double bond is less than a single bond but more than half the length. The double bond consists of 1 σ bond and 1 π bond.

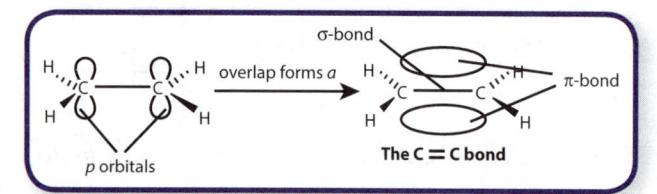

Shape

Three groups of electrons around the carbons of the double bond give a bond angle of approximately 120° and planar shape.

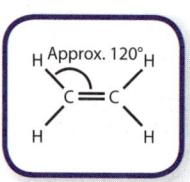

Stereoisomers

Alkenes can show geometric isomerism (also known as E/Z isomerism or cis-trans isomerism). Isomers have the *same structural formula but a different arrangement of atoms in space*. They arise when there is restricted rotation about a double bond (double bond provides this) **and** there are two different groups on each carbon on either side of the double bond.

Example
1-chloro-2-methylbut-1-ene shows E/Z isomerism.

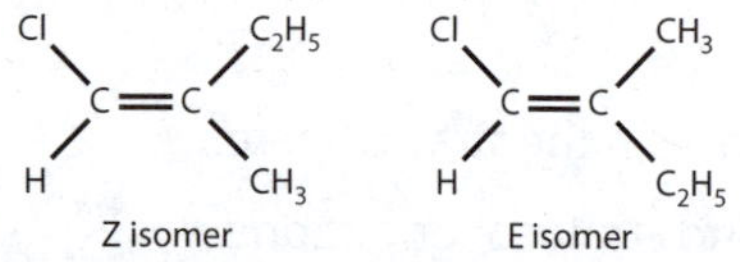

Z isomer   E isomer

Cis and trans isomers are a special type of E/Z isomer. If two of the same group are on the **same** side of the double bond the molecule is the **cis** isomer. If there are two of the same group on **different** sides of the double bond the molecule is the **trans** isomer.

E/Z Isomers

To determine which is E or Z, consider each carbon of the double bond separately. Put the two groups attached to the first carbon of the double bond in order of priority using the **Cahn-Ingold-Prelog** (CIP) rules below.

1. Check the first atom of the two groups attached to C of the double bond. Highest priority goes to the atom with the highest atomic mass (including the highest mass isotope if relevant).

2. ● If both first atoms have the same atomic mass check the mass of the atoms bonded to this first atom. Write them in order of decreasing atomic mass. If these are the same, keep going out atom by atom until there is a difference between the two groups.

   ● If there is a double bond then write this as two atoms, e.g. in an aldehyde group, H–C=O, count atoms attached as O O H.

Repeat for the second carbon of the double bond.

If the two highest priority groups for both carbons are on the same side of the double bond the molecule is the Z isomer. If they are on opposite sides of the double bond the molecule is the E isomer.

Reactions of Alkenes

Electrophile: Species that can receive a pair of electrons to form a covalent bond to carbon.

Alkenes undergo **electrophilic addition** reactions. The double bond breaks open and two new groups attach to the carbon at either end.

Alkene to bromoalkane: Mix with HBr at room temperature.

Step 1: The H—Br bond breaks. H$^+$ receives a pair of electrons from the double bond to form a covalent bond to the carbon with most H attached (this is known as Markovnikov's rule). The carbon on the other side of the double bond forms a **carbocation** (a positively charged carbon atom).

Step 2: Br$^-$ donates a pair of electrons to the carbocation.

The electrophile can attach at either side of the double bond to form the most stable carbocation. A tertiary cation is the most stable, followed by a secondary; a primary cation is least stable.

Alkene to dibromoalkane: Mix with Br$_2$ in an organic solvent, e.g. hexane, at room temperature.

$$CH_2CH_2 + Br_2(org) \longrightarrow CH_2BrCH_2Br$$

If the bromine is used aqueous Br$_2$(aq) the product formed is CH$_2$(OH)CH$_2$Br. This provides evidence that the reaction proceeds via a carbocation intermediate.

The Br$^-$ and H$_2$O compete to donate a pair of electrons to the carbocation. Since there are many more water molecules than bromide ions the main product is the bromoalcohol.

Test for alkenes: An aqueous solution of bromine turns from brown to colourless.

Alkene to alcohol: Heat 330°C, 6 MPa with steam and H$_3$PO$_4$ catalyst.

To form a diol, mix with KMnO$_4$, an oxidising agent. An oxidising agent can be simplified to [O].

$$CH_2CH_2 + H_2O + [O] \longrightarrow CH_2(OH)CH_2OH$$

Alkene to alkane: Heat to 150°C at 5 atm. with H$_2$ and Ni catalyst or with a Pt catalyst at room temperature and pressure.

DAY 6

This reaction is used in hydrogenation of vegetable oils to make margarine. The quantity of H_2 is controlled so that only some of the double bonds are hydrogenated: the greater the hydrogenation, the higher the melting point.

Alkene to addition polymer: Heat to 200°C at 1500 atm. with a catalyst.

Polymerisation: The chemical reaction where many small organic molecules (monomers) join to form a very long macromolecule (polymer).

Drawing Polymers

To draw the polymer given the monomer: Draw the monomer with the double bond horizontal and all attached groups at approximately 120°.

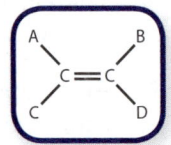

Move the attached groups to a horizontal position, remove one of the bonds of the double bond and draw new bonds from the carbon at each end of the double bond to a new monomer.

The bond joining a monomer is always between the carbon atoms of the double bond.

To draw the monomer given the addition polymer chain: Look for a repeat in the polymer. The carbon next to the repeat has a double bond in the monomer.

Polymer Properties

Addition polymers are unreactive because the double bond of the monomer has been lost.

The properties of a polymer depend on the intermolecular bonding between polymer chains. Plasticisers can be added to polymers as they form. These act as lubricants between the polymer chains by interfering with the intermolecular bonds and allowing the polymer chains to slide over each other more easily.

PVC is polyvinylchloride, systematic name poly(chloroethene).

A plasticiser is added to make the polymer more flexible for use as imitation leather, electrical wiring insulation and similar purposes.

uPVC is poly(chlorethene), which is unplasticised. This is hard and rigid and used for window frames, drainpipes and guttering.

QUICK TEST

1. Draw the skeletal structure of 2,5-heptadiene.
2. State the two types of covalent bond in an alkene.
3. Explain the difference between a saturated and an unsaturated molecule.
4. Describe a test for alkenes.
5. State the reaction conditions for the conversion of propene to propanol.
6. Draw the repeating unit for poly(phenylethene).

PRACTICE QUESTIONS

1. A polymer can be formed from the monomer shown.

   What is the name of the polymer? [1 mark]

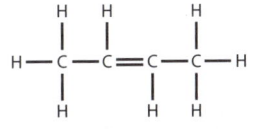

   A Poly(1,2-dimethylethene)   B Poly(1,2-dimethylethane)

   C Poly(but-2-ene)   D Poly(but-2-ane)

2. Which of the following molecules is a Z isomer? [1 mark]

   A   B   C   D

3. During the manufacture of margarine, vegetable oils are converted into solid fats. This is done so that the low melting point unsaturated oils will become higher melting point saturated compounds. One concern about the process is that it results in the production of some **trans** fats, which contain E isomers. These are thought to be less healthy than the naturally occurring **cis** fats, which contain only Z isomers.
   a) Explain what is meant by unsaturated. [1 mark]
   b) Describe a test that could be used to show that a vegetable oil is unsaturated. [2 marks]
   c) Name the conditions required for the conversion of saturated to unsaturated molecules. [2 marks]
   d) Name the reaction mechanism for this reaction. [2 marks]
   e) Explain how unsaturated fats show E/Z isomerism. [2 marks]

4. A haloalkane is produced when pent-1-ene is reacted with concentrated hydrochloric acid.
   a) Use molecular formula to write a balanced equation for this reaction. [2 marks]
   b) There are two possible products for this reaction. Use displayed formulae to show and name both products. [4 marks]
   c) Explain why one of the products listed in part b) is more likely to be produced than the other. [4 marks]

5. Addition polymers can be made from alkenes. The properties of the polymer depend on a number of factors including the monomer chosen and the reaction conditions of the polymerisation.
   a) Draw the displayed formula of the monomer from which poly(chloroethene) is made. [2 marks]
   b) Draw a length of the polymer chain for poly(tetrafluoroethene). [2 marks]
   c) Identify the monomer that was used to make the polymer shown below. [2 marks]

   d) Give the reaction conditions needed for the polymerisation of ethene. [1 mark]

99

DAY 7 — 60 Minutes

Alcohols

Naming

Alcohols are named in the same way as alkanes but with the suffix -ol. Position and number of —OH groups is indicated by the number of the carbon atom to which they are attached and use of di, tri etc.

Example
3-methylbutan-1,2-diol

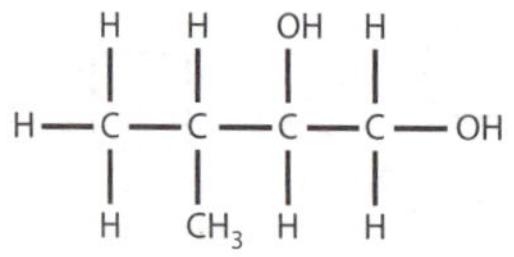

Note the —OH group takes preference over alkyl groups when numbering the carbon chain, so the alcohol group has the lowest number.

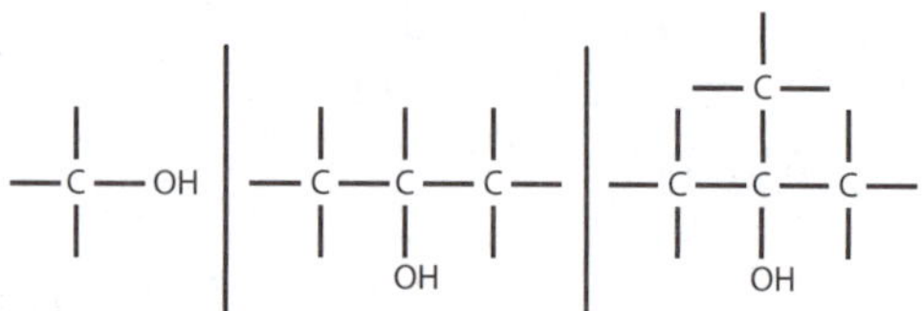

Primary | Secondary | Tertiary

Primary alcohols have the alcohol group attached to the end of a carbon chain, i.e. the C bonded to 2 hydrogens: RCH_2OH.

Example
Butan-1-ol

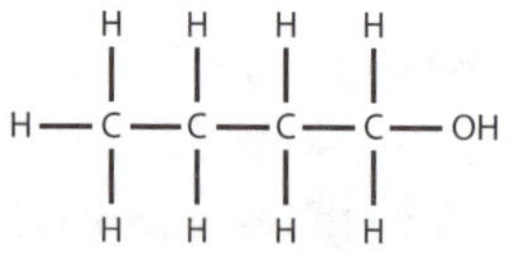

Secondary alcohols have the alcohol group attached to a C in the chain, i.e. a carbon bonded to 1 hydrogen and 2 R groups: R_2CHOH.

Example
Butan-2-ol

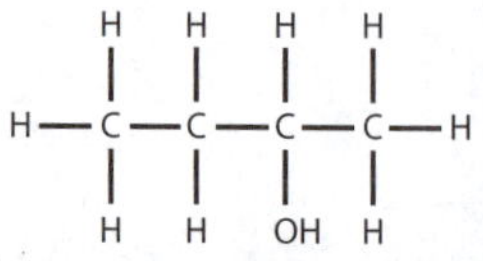

Tertiary alcohols have the alcohol group attached to a C that is not bonded to hydrogen: R_3COH.

Example
2-methylpropan-2-ol

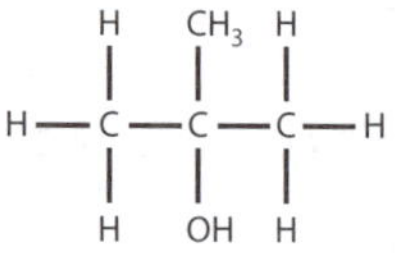

Bonding

The C—O—H bond angle is about 104.5°. On the oxygen atom of the functional group there are two lone pairs of electrons as well as two bonding pairs of electrons.

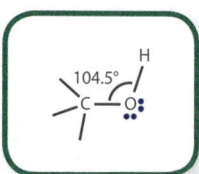

The O—H bond is polarised because of the large difference in electronegativity between the oxygen and hydrogen atoms in the functional group. So alcohol groups can form hydrogen bonds between other alcohol molecules as well as other chemicals like water.

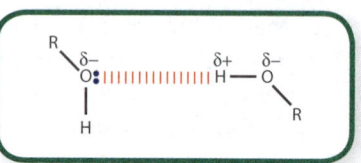

This means alcohols have a higher boiling point than alkanes of a similar molecular mass.

Propan-1-ol M_r 60   b.pt 97°C

Butane M_r 58   b.pt −1°C

Hydrogen bonding means that alcohols are water soluble providing the R group is not too large. The first three alcohols are significantly water soluble.

Alcohols as Solvents

Alcohols make useful solvents because they are able to dissolve some substances that are insoluble in water. A mixture of ethanol and water is used as a solvent for reactions of haloalkanes and allows aqueous reagents to mix with the haloalkane.

Reactions of Alcohols

Alcohol to alkene: Reflux with concentrated sulfuric acid.

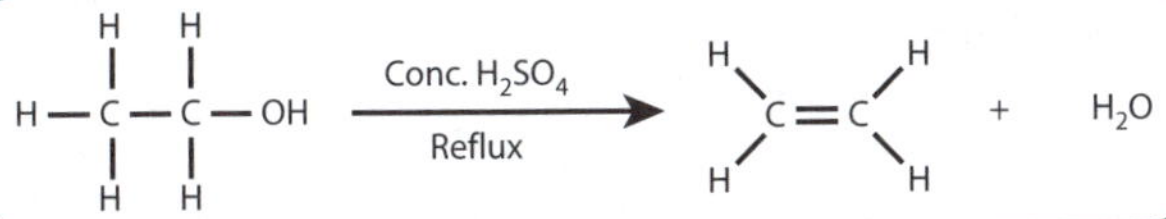

This is known as a dehydration reaction because water is removed.

The type of mechanism is **elimination**.

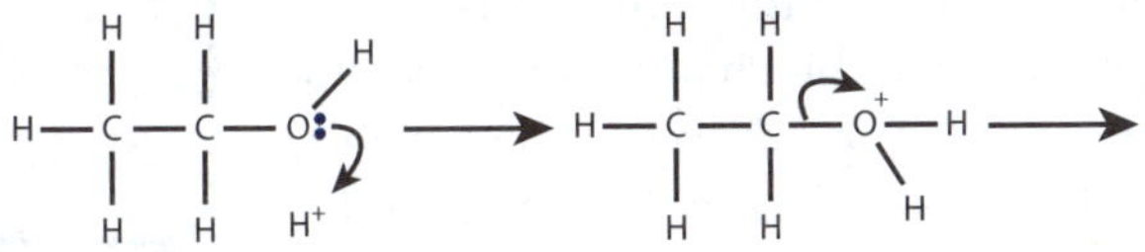

The —OH group gains a proton from the acid giving a positive charge on the oxygen atom. The C—O bond breaks eliminating H₂O and leaving a carbocation.

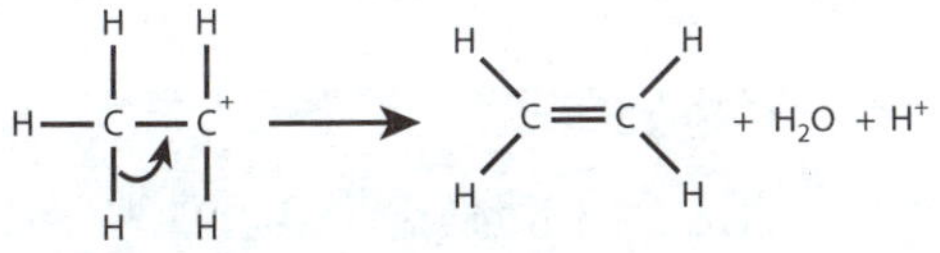

Electrons from the neighbouring C—H bond transfer to form a double bond.

Secondary and tertiary alcohols can form one of two alkene isomers. Which isomer forms is determined by which H is eliminated.

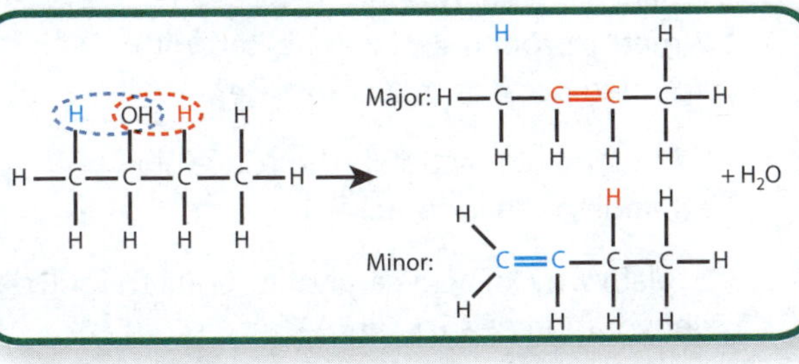

When the alkene formed has two different groups attached to each of the carbons on either side of the double bond, the product shows E/Z isomerisation (see page 96). This means that three isomers are possible as product.

E.g. for elimination from propan-2-ol, the two isomers are the ones shown on this page and this E isomer:

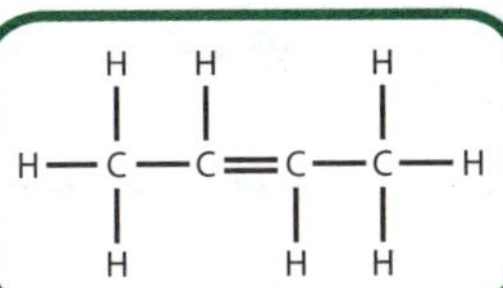

Alcohols can also be dehydrated by passing alcohol vapour over a heated Al_2O_3 catalyst.

Alcohol to haloalkane: nucleophilic substitution.

Chloroalkanes – add phosphorus(V) chloride.

$CH_3CH_2OH(l) + PCl_5(s) \longrightarrow CH_3CH_2Cl(g) + POCl_3(l) + HCl(g)$

Bromoalkanes – add 50% H_2SO_4 and potassium bromide to produce HBr in situ.

$CH_3CH_2OH(l) + HBr(g) \longrightarrow CH_3CH_2Br(l) + H_2O(l)$

Iodoalkanes – add red phosphorus and iodine to form PI_3 in situ.

$3CH_3CH_2OH(l) + PI_3 \longrightarrow 3CH_3CH_2I(l) + H_3PO_3(l)$

Alcohol to aldehyde, ketone or carboxylic acid: Oxidation of primary and secondary alcohols results in a different functional group.

Oxidation is carried out by heating with acidified postassium dichromate ($H^+/K_2Cr_2O_7$).

Primary alcohols:

Step 1

$CH_3CH_2OH(aq) + [O] \longrightarrow CH_3CHO(aq) + H_2O$
aldehyde

This can then further oxidise.

Step 2

$CH_3CHO(aq) + [O] \longrightarrow CH_3COOH$
carboxylic acid

To remove the aldehyde at the end of step 1, heat under distillation.

Secondary alcohols:

$CH_3CH(OH)CH_3 + [O] \longrightarrow CH_3COCH_3 + H_2O$
ketone

No further oxidation can occur.

Tertiary alcohols cannot be oxidised with potassium dichromate.

During oxidation orange dichromate is reduced to green chromium ions.

Maybe find flash cards about naming and functional groups

DAY 7

To make an ester heat an alcohol with a carboxylic acid in the presence of a concentrated acid catalyst.

Combustion of Ethanol

Alcohols burn to form carbon dioxide and water.

$$CH_3CH_2OH(l) + 3O_2(g) \longrightarrow 2CO_2(g) + 3H_2O(g)$$

They burn very 'cleanly' because of the oxygen already present in the molecule. There is little incomplete combustion.

Further Functional Groups Containing Oxygen (OCR B only)

Ethers contain the functional group C—O—C.

Example
Ethoxyethane

They are functional group isomers of alcohols.

Carboxylic acids contain the functional group COOH.

Example
Propanoic acid

Acid anhydrides contain this functional group:

Example
Ethanoic anhydride

Esters contain this functional group:

Example
Ethylpropanoate

Phenols

Phenols have an alcohol group attached directly to one of the carbons in a benzene ring. This changes the nature of the —OH group, which becomes slightly acidic. The —OH group in phenols will react with alkalis to form a salt but is not acidic enough to react with carbonates to release carbon dioxide.

Phenols will not react with carboxylic acids to form esters but will form esters with acid anhydrides.

QUICK TEST

1. Draw the structure of propan-1,2,3-triol and state whether it is a primary, secondary or tertiary alcohol.

2. State the reaction conditions required to make propene from propan-2-ol.

3. What colour change would be seen if 2-methylpropan-2-ol was heated with acidified potassium dichromate?

4. What type of reaction is the conversion of ethanol to ethanoic acid?

5. Explain why an acid catalyst is required for the dehydration of an alcohol.

6. Explain why short chain alcohols are water soluble.

7. Write a balanced equation to show the formation of chloroethane from ethanol.

102

PRACTICE QUESTIONS

1. The conversion of butanol to butene is an example of [1 mark]

   A elimination and hydrolysis   □   B elimination and dehydration □

   C addition and hydrolysis   □   D addition and dehydration.   □

2. Which of the following would cause a colour change if heated with acidified potassium dichromate? [1 mark]

   i. Methylpropan-1-ol   ii. Methlypropan-2-ol

   iii. 3-ethylpentan-3-ol   iv. 3-ethylpentan-2-ol

   A **i** and **ii**   □   B **ii** and **iii**   □

   C **ii**, **iii** and **iv** □   D **i** and **iv**.   □

3. Which of the following is true of phenols? [1 mark]

A	They are more acidic than alcohols.
B	They react with carboxylic acids to form esters.
C	They react with carbonates to release carbon dioxide.
D	They are aliphatic compounds.

4. a) Draw the displayed formula of a primary, a secondary and a tertiary alcohol with the molecular formula $C_4H_{10}O$. [3 marks]

   b) Give the IUPAC name for each of your structures. [3 marks]

   c) For each of your structures draw the displayed formula for the product of its reaction with excess acidified postassium dichromate. [3 marks]

5. Ethene can be produced from ethanol.

   a) State the reagents and conditions for this reaction. [2 marks]

   b) Draw the displayed formula of the reactants and products and use curly arrows to show the mechanism for this reaction. [4 marks]

   c) Explain why a mixture of products is obtained if the same reaction is carried out using butan-2-ol in place of ethanol. [2 marks]

6. Complete the table.

Reactant	Reaction conditions	Product	Type of reaction
ethanol		ethanoic acid	
	pass the vapour over heated Al_2O_3	propene	
butan-1-ol	PCl_5 or concentrated HCl		

[6 marks]

DAY 7 — 60 Minutes

Experimental Techniques

Refluxing

Refluxing allows reactants to be heated together at boiling point without loss of reactants, products or solvent.

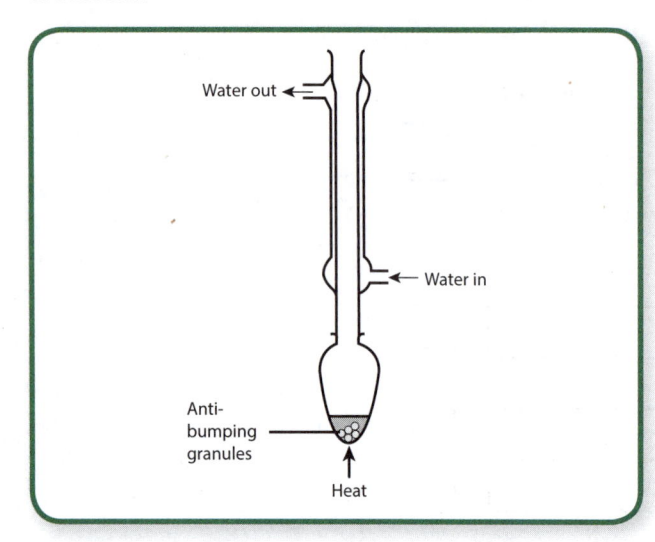

The condenser is vertical so that the condensing liquid falls back into the reaction mixture.

- The top of the condenser must be open to prevent pressure build up.
- Water goes in at the bottom of the condenser and out at the top to prevent air being trapped inside the outer jacket of the condenser, which would result in inefficient cooling.
- The joint between the flask and the condenser must be tight to prevent gases escaping.
- The heating should be adjusted so the liquid drips into the flask at a slow steady rate and reactants do not froth up into the condenser.
- Inert anti-bumping granules are added to the flask to provide a surface on which many small bubbles can form. This prevents sudden large bubbles blowing up into the condenser.

Distillation

Distillation is used for separating organic liquids with different boiling points or to separate a solvent from solutes.

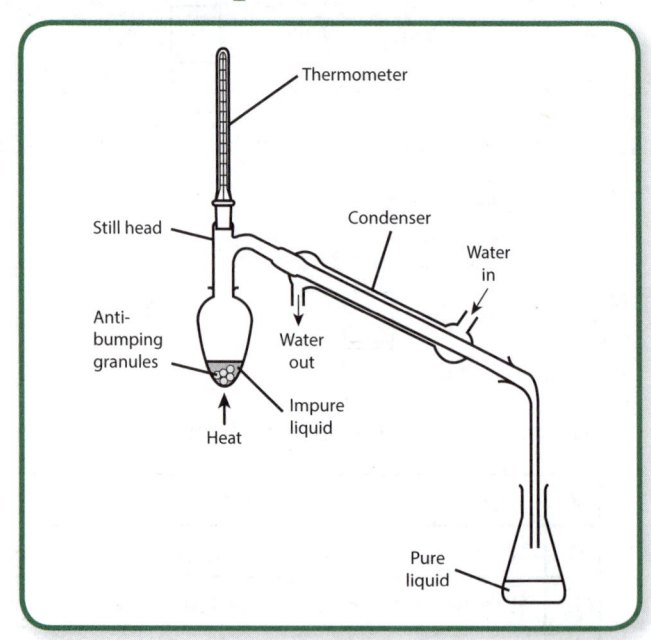

The temperature in the flask rises until it reaches the boiling point of one of the liquids in the mixture. This vaporises and passes into the condenser, where it cools, condenses and runs into the collecting vessel. The temperature remains constant until all of this component (fraction) has vaporised. The thermometer is placed with the bulb opposite the opening to the condenser (the side arm of the still head) so that the temperature of the vapour can be monitored. Once the temperature shoots up, the collecting vessel is changed. The temperature in the flask rises until the second component vaporises and is collected in the same way.

- The top of the still head must be closed off with the thermometer.
- The end of the condenser is attached to a delivery tube, which must be open at the end to allow the condensate to escape and to prevent pressure build up.
- All other joints must be tightly sealed.
- Water goes in at the lowest point in the condenser and comes out at the highest point. This sets up a counter-current where the water in the condenser is always colder than the temperature of the vapour so reduces the chance of vapour escaping from the delivery tube.

104

immiscible - not forming homogenous mixture

- An electric heating mantle is the best way of heating organic liquids. It allows the rate of heating to be easily controlled and reduces the risk of fire compared to a naked flame.

Separating Immiscible Liquids

A separating funnel allows immiscible liquids to be shaken together for reacting and then separated out. Both liquids are poured into the funnel with the tap closed. The funnel is stoppered and the liquids mixed by rocking the funnel. Any pressure build up in the funnel can be released by opening the tap when the funnel is inverted. The funnel is placed vertically in a clamp and the liquids are allowed to separate into two layers. A beaker is then placed below the tap, which is opened to allow the bottom layer to run into the beaker until the meniscus between the two layers enters the tap.

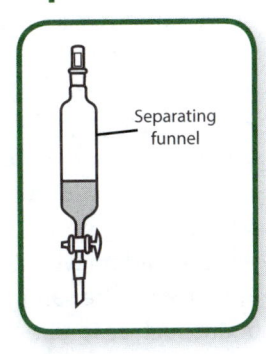

Separating funnel

Example
To prepare a haloalkane from an alcohol

Mix the reactants:
Put concentrated HCl(aq) and the alcohol in the funnel with the tap closed. Stopper the funnel and gently rock to allow the organic and aqueous layers to mix and the reaction to occur. Let the two layers settle out. The organic layer is typically less dense and will float on the top of the aqueous layer. Open the tap, let the aqueous layer run out into a beaker and discard it.

Remove excess acid:
Add sodium carbonate solution to the funnel. Stopper and gently shake. Any remaining acid reacts with the carbonate to produce carbon dioxide so the stopper must be removed periodically to release the pressure. Run off the aqueous layer. Repeat this until there is no further fizzing when the carbonate is added. Finally, run the organic product into a separate container.

Remove remaining water:
If the organic product looks cloudy this shows that it contains water. A solid drying agent can be added, which absorbs the water. Suitable drying agents are anhydrous sodium sulfate or anhydrous calcium chloride. Add a few grams of the drying agent to the organic product and swirl until the liquid is completely transparent. Separate the product from the drying agent by decanting, i.e. pour off the liquid leaving the solid behind.

Purify the product:
Distil the mixture of products and collect the fraction that boils at the known boiling point of the desired product.

Recrystallisation

Recrystallisation is used to purify an organic solid.

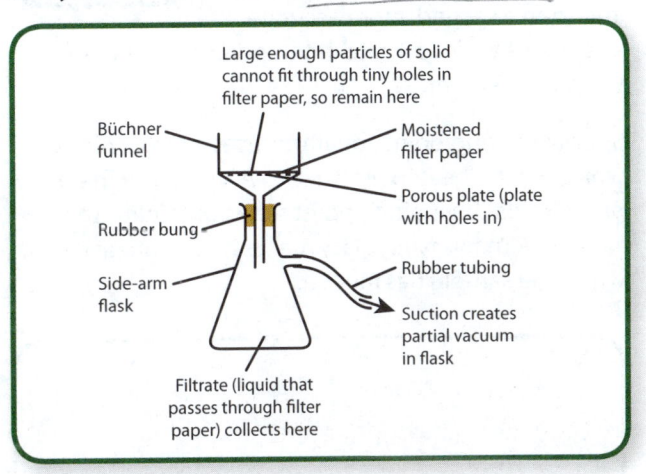

Büchner funnel — Large enough particles of solid cannot fit through tiny holes in filter paper, so remain here
Moistened filter paper
Porous plate (plate with holes in)
Rubber bung
Rubber tubing
Side-arm flask
Suction creates partial vacuum in flask
Filtrate (liquid that passes through filter paper) collects here

The impure sample is dissolved in the **minimum** amount of **warm** solvent. Any insoluble impurities are removed by decanting or filtering once all the sample has dissolved. The solution is then allowed to cool. New pure crystals form. Any soluble impurities remain in solution. The crystals are filtered under vacuum so that the solvent and dissolved impurities are removed. The crystals are washed on the filter paper with a minimum of cold solvent to remove surface impurities without dissolving the crystals. They are dried in a **cool** oven to prevent any degradation (breakdown) of the compound.

[The solvent must be carefully selected so that the substance to be purified is more soluble in hot solvent than in cold.]

105

DAY 7

Finding a Melting Point

The melting point of a substance can be used to help with identification and determine purity. A pure substance will melt over a narrow range of temperatures (no wider than 2°C) at the expected melting point.

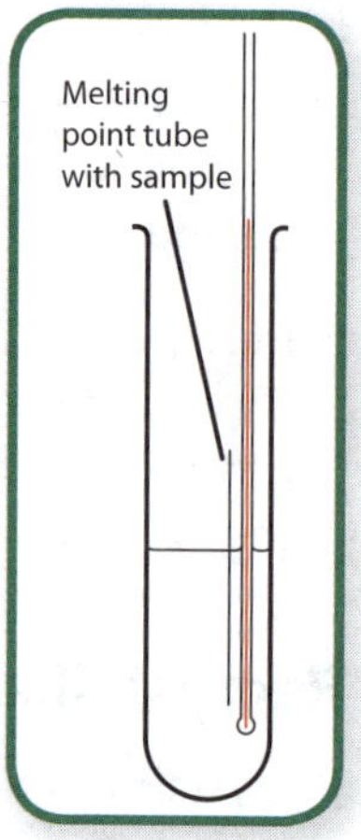

A dry sample is crushed to powder and put into a melting point tube, which has been sealed at one end. The tube is attached to a thermometer, which is gently heated in a boiling tube containing oil. As soon as the first signs of melting are seen the temperature is noted. It is noted again when all the sample has melted. When the solid melts it changes from opaque to transparent.

A first rough measurement is made and then a second, more accurate one with a very slow rate of temperature rise close to the melting point.

Special melting point apparatus uses a metal block, which can be heated electrically. A slot in the metal block holds the melting point tube containing the sample. A magnifying glass makes it easier to see when the sample has melted.

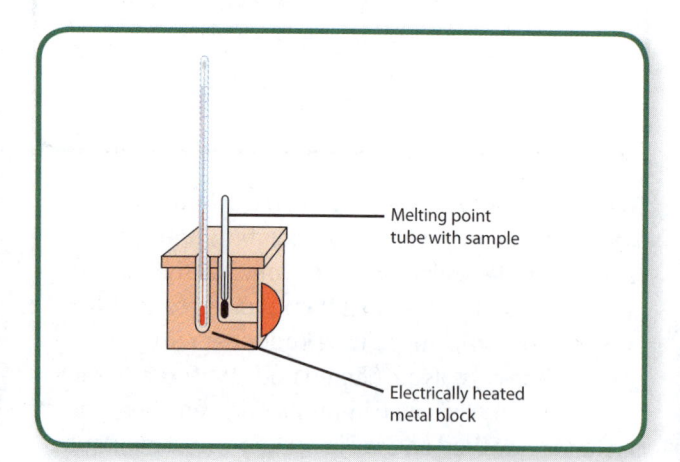

Thin-Layer Chromatography (TLC)

TLC can be used to separate out dissolved substances.

The test substance is dissolved and spotted onto a thin layer plate about 1 cm from the bottom using a capillary tube. The spot is kept to as small an area as possible. The plate is placed in a developing tank containing solvent that is below the level of the sample spot. The tank could be made from a beaker covered with a watch glass to allow a saturated atmosphere of solvent to form. The solvent is allowed to rise up the plate by capillary action. When the solvent is near the top, the solvent front is marked with a pencil. The plate is removed and dried. If the sample is not coloured, a locating agent such as iodine vapour is used to find the sample spots by staining them. The plate is placed in a beaker containing a few crystals of iodine. A pure sample will produce only one spot.

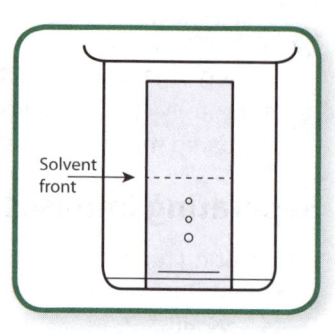

QUICK TEST

1. What is the purpose of refluxing?
2. Where does the cooling water enter the condenser?
3. Suggest how an organic liquid can be dried.
4. How can you tell which is the organic layer in two immiscible liquids?
5. Suggest how you could purify an organic solid.
6. Describe how you can identify a pure sample using melting point.

PRACTICE QUESTIONS

1. A student synthesised a sample of aspirin. Which of the following results indicate that the sample was impure? The accepted melting point of aspirin is 136°C. **[1 mark]**

   A The melting point range of the sample was 134.5–136.0°C. ☐
   B There was no colour change with neutral iron(III) chloride. ☐
   C Thin-layer chromatography gave two spots. ☐
   D The sample was pure white. ☐

2. A student prepared a sample of 2-methyl-2-chloropropane. The melting point is −20°C and the boiling point is 51°C. A suitable way to purify the product is **[1 mark]**

   A distillation ☐
   B chromatography ☐
   C recrystallisation ☐
   D filtering. ☐

3. Butanoic acid can be prepared by boiling butan-1-ol with acidified potassium dichromate for 15 minutes.

   a) Draw a labelled diagram of the apparatus you would use to boil the reactants. **[3 marks]**

   b) Once the reaction is complete, the pure butanoic acid can be separated from other substances by distillation. A student drew a diagram of distillation apparatus. Describe two errors in the diagram. **[2 marks]**

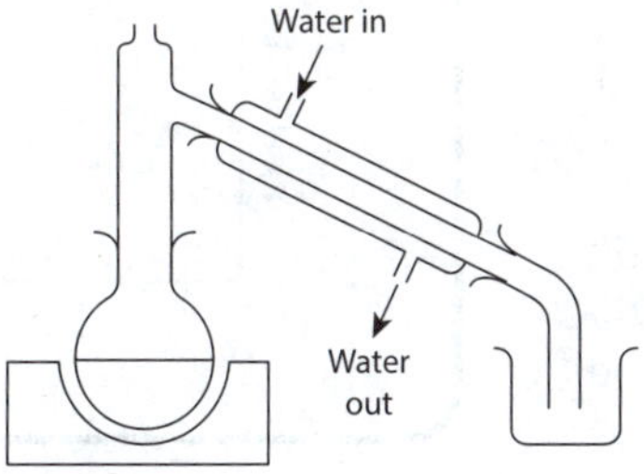

   c) Suggest a reason why the student chose to use an electric heating mantle rather than a Bunsen burner to heat the mixture. **[1 mark]**

4. a) Describe the process of purifying an organic solid by recrystallisation. **[5 marks]**

   b) How does taking the melting point of the purified solid help to determine the purity? **[2 marks]**

DAY 7 — 60 Minutes

Mass Spectrometry

Mass spectrometry can be used to calculate relative atomic mass of elements and to help to determine the structure of organic molecules.

Principles of a *Time of Flight* TOF Spectrometer

Simplified Diagram of a Time of Flight Spectrometer

(Diagram labels: Sample inlet, Ionisation area, Acceleration area, Light ions, Heavy ions, Ion-detector, Drift zone, Time measurement, Vacuum chamber)

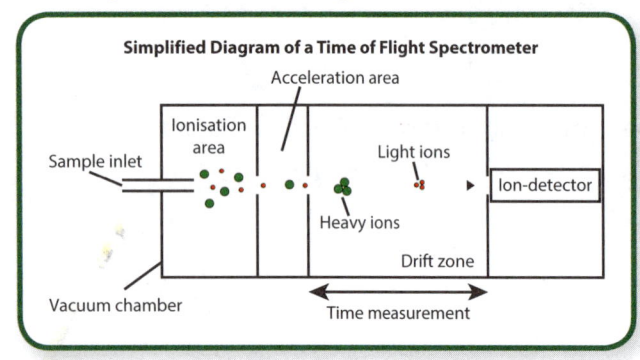

1. The sample is ionised, for example by passing through a high energy beam of electrons. Molecules are broken into fragments and some have electrons knocked off.

2. Positive ions are accelerated in an electric field so that all ions have the same kinetic energy. The higher the mass of the ion, the longer the time taken for the ions to reach the detector.

3. Accelerated ions move through the drift zone arriving at a detector at different times according to their mass.

4. The time of flight is converted into mass ÷ charge (m/z) and a spectrum is drawn showing the relative quantity of each ion at each mass. (Charge can be assumed to be 1+ for A Level therefore m/z = mass of fragment.)

Determining A_r of an Element

The spectrum gives a peak for each isotope of the element. The height of the peak is the relative abundance of the isotope.

E.g. Copper
Peak at 63 relative abundance 70
Peak at 65 relative abundance 30

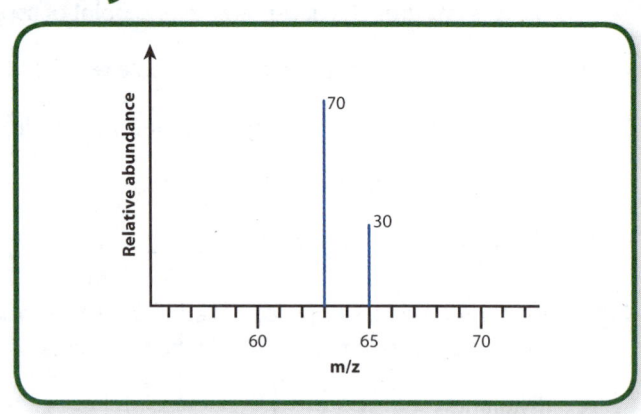

(Copper mass spectrum: peak at 63 abundance 70, peak at 65 abundance 30)

Multiply relative abundance of each peak by the mass of the peak. Add all values together and divide by 100. E.g. for the copper spectrum above:

weighted average

$$\frac{(63 \times 70) + (65 \times 30)}{100}$$

$A_r = 63.6$

Elements that form diatomic molecules form more complex spectra, e.g. Cl_2.

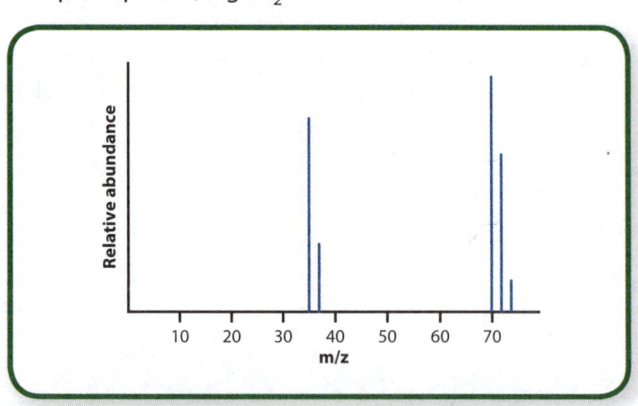

(Chlorine mass spectrum with peaks around m/z 35, 37 and 70, 72, 74)

m/z	Ion giving peak
35	$^{35}Cl^+$
37	$^{37}Cl^+$
70	$[^{35}Cl-^{35}Cl]^+$
72	$[^{35}Cl-^{37}Cl]^+$
74	$[^{37}Cl-^{37}Cl]^+$

Getting Information from Complex Spectra

Organic molecules break into many fragments and have complex spectra. It is not necessary to understand every peak.

Deciding the M_r of a Molecule

The peak with the highest m/z is the molecular ion. This is the entire molecule with one electron missing. The m/z value of the molecular ion peak is the M_r of the molecule. For the unknown molecule, X, in the spectrum below, this is 72.

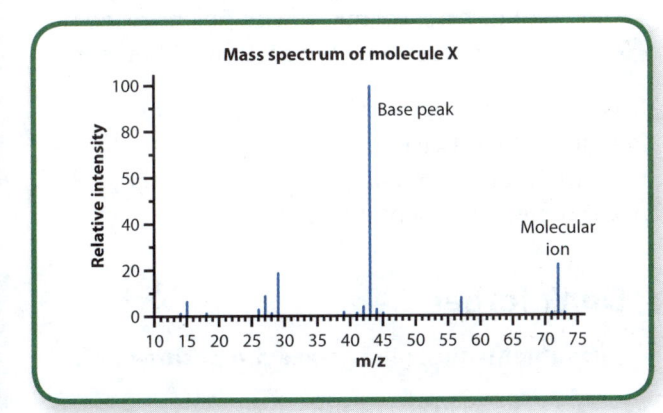

Mass spectrum of molecule X

(Base peak at m/z ~43; Molecular ion at m/z 72)

M+1 Peak

A very small peak at m/z one larger than the largest significant peak is the M+1 peak. It arises because some of the molecular ions will contain a ^{13}C atom. The more carbon atoms in the molecule, the higher the probability a molecule will have a ^{13}C atom and the larger the M+1 peak.

Deciding the Formula of the Molecule

High resolution mass spectrometers can be used to calculate the molecular mass of a molecule to 4 decimal places. This can be used together with accurate atomic masses to find the formula of the molecule.

Examples of accurate atomic masses:

H = 1.0078

C = 12.0000

O = 15.9949

Example

A low resolution mass spectrum gives a molecular ion with m/z of 44. Possible molecular formula include C_3H_8, C_2H_4O and CO_2.

If the high resolution spectrum gives M_r 44.0261, this eliminates two of the possible matches.

$C_3H_8 = 44.0624$

$C_2H_4O = 44.0261$

$CO_2 = 43.9898$

The only match is C_2H_4O.

Deciding the Structure of the Molecule

The fragmentation pattern allows evidence to be gathered to propose the structure of the molecule. The electron beam causes random bond fission to give a positive fragment ion and a radical, e.g.

$$CH_3CH_2CH_3 \longrightarrow [CH_3]^+ + \bullet CH_2CH_3$$

Peaks in the spectra represent the positive ions.

The difference in mass between the molecular ion and the fragment ion peaks = the mass of the radical that was broken off.

The peak from the most abundant fragment is known as the **base peak** and is set as 100% intensity.

Note: Unlike the mass spectrum of elements, the peaks do not all add up to 100%.

Commonly Occurring Fragment Ions

m/z	Possible ion
15	$[CH_3]^+$
29	$[CH_3CH_2]^+$
43	$[CH_3CH_2CH_2]^+$ $[CH_3CO]^+$
77	$[C_6H_5]^+$

To propose a molecular structure from a spectrum:

1. Determine the M_r from the highest m/z peak and propose a molecular formula from this and other data.

DAY 7

For molecule X (see page 109), $M_r = 72$.

Assuming that additional information (such as an IR spectrum or chemical analysis) suggests that a carbonyl group is present.

Possible formula = C_4H_8O

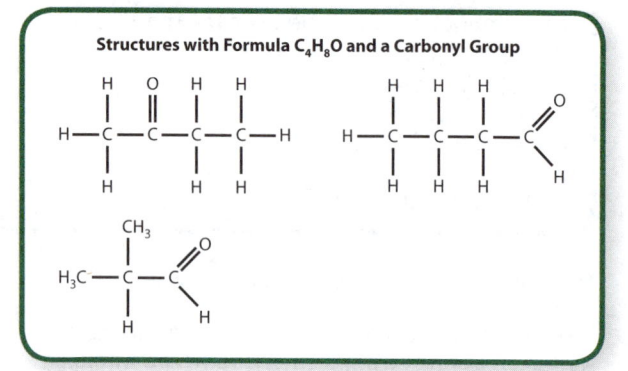

Structures with Formula C_4H_8O and a Carbonyl Group

2. Identify major peaks in the spectrum and suggest fragment ions that have produced them.

Major peaks for molecule X (see page 109):

m/z	Possible ions
15	$[CH_3]^+$
29	$[CH_3CH_2]^+$
43	$[CH_3CH_2CH_2]^+$ $[CH_3CO]^+$
57	$[CH_3CH_2CO]^+$

3. Fit the fragments together to propose structures and use the spectrum and any other data given to select the most likely isomer.

Mass of likely fragments are all seen in the spectrum.

To confirm this is correct more information is needed. For example:

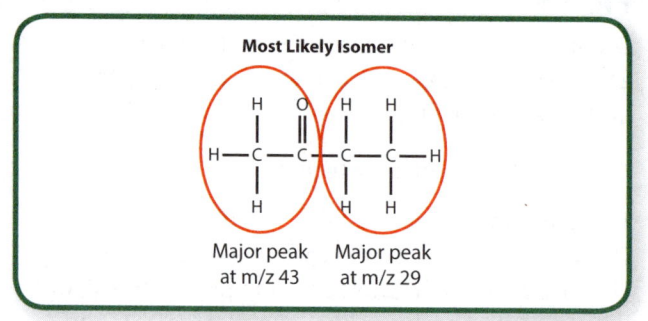

Most Likely Isomer

Major peak at m/z 43   Major peak at m/z 29

- the molecule does not contain an aldehyde group, does not give a silver mirror with Tollens' reagent
- the spectrum matches the spectrum of butanone.

Mass spectra can be used like fingerprints for identification. This allows positional isomers with the same molecular formula to be identified as they will have different fragmentation patterns.

Don't forget

...fragments that give a peak are positive ions.

$[CH_3]^+$ not CH_3

...fragments that have been lost are radicals.

•CH_3

QUICK TEST

1. In a TOF spectrometer, which ion would reach the detector first, $[^{79}Br]^+$ or $[^{81}Br]^+$?
2. How are the ions formed in a mass spectrometer?
3. How many peaks would be seen in the mass spectrum of Br_2 given that it has only two isotopes?
4. Give the expected value of m/z for the molecular ion peak and M+1 peak of hexan-1-ol.
5. Suggest a possible formula for a mass spectrum peak at m/z 77.
6. What fragment is missing from the whole molecule if the molecular ion has m/z 46 and the next highest peak has m/z 31?

PRACTICE QUESTIONS

1. The mass spectrum of lithium shows two peaks: m/z 6 is 7.6% and m/z 7 is 92.4%. The A_r for lithium is [1 mark]

   A 6.9

   B 6.5

   C not possible to calculate

   D 7.0

2. The accurate molecular mass of a compound is 28.0312. Given the following values
   H = 1.0078   C = 12.0000   O = 15.9949   N = 14.0031
   which is the formula of the compound? [1 mark]

   A CO

   B N_2

   C C_2H_4

   D HCN

3. Zirconium is an element. The mass spectrum of zirconium is shown below.

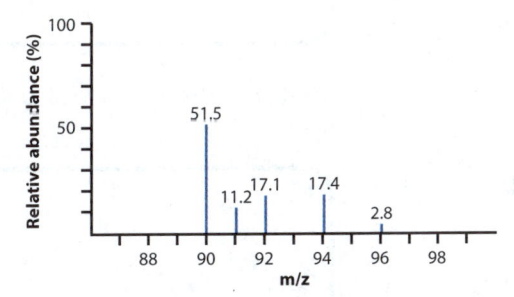

   a) Write the formula of the species making the peak at 90. [2 marks]

   b) Describe how the different fragments are separated in a TOF spectrometer. [3 marks]

   c) Calculate the relative atomic mass of zirconium. [1 mark]

4. Gallium has a relative atomic mass of 69.7. It has two isotopes, gallium-69 and gallium-71.
   Sketch a diagram of the mass spectrum you would expect to see for a typical sample of gallium. [2 marks]

5. A molecule with formula C_3H_8O has a mass spectrum with the major fragments shown in the table to the right.
   The molecule was shown to contain an alcohol group.

	m/z
	29
Base peak	31
	59
Molecular ion peak	60

   a) Suggest the formula of the molecular ion and base peak fragments. [3 marks]

   b) Suggest the fragment that was removed from the whole molecule to give the peak at 31. [1 mark]

   c) Suggest a structure for the molecule. [2 marks]

   d) How could your structure be confirmed? [1 mark]

DAY 7 — 60 Minutes

Infrared Spectroscopy

Covalent bonds are shared pairs of electrons that are vibrating. When **infrared energy** is absorbed by some bonds it causes them to vibrate **more**.

Identifying Unknown Molecules

An infrared spectrum shows the absorption of energy at different wavenumbers. An absorption is shown as a trough, which is known as a peak! Each different molecule has a characteristic absorption pattern that can be used to identify the molecule by comparing to a database of standard spectra.

Measuring the Concentration of Known Molecules

The intensity of absorption (depth of the peak) at a specific wavenumber for a particular molecule is proportional to the concentration of that molecule.

Infrared absorption can be used to measure the concentration of gases, for example: the concentration of carbon dioxide in the atmosphere; the concentrations of carbon monoxide and nitrogen oxide emissions from vehicles; the quantity of alcohol in the breath as an estimate of blood alcohol content.

Using IR to Identify Bonds and Functional Groups

Different covalent bonds absorb at different wavenumbers of infrared energy. The absorbance values will be provided in the exam. (One bond may absorb at more than one wavenumber.) Many bonds absorb at wavenumbers below 1500 cm^{-1} so the absorption peaks overlap making them difficult to pick out individually. This region of an IR spectrum is known as the fingerprint region.

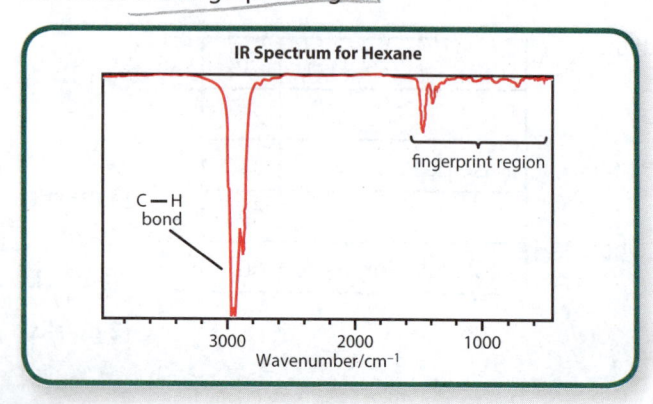

IR Spectrum for Hexane

The C—H bond gives a peak at around 2850–3100 cm^{-1}. It is not very useful for identification since it is present in almost all molecules.

Bonds in some functional groups produce characteristic absorptions. The exact wavenumber of absorption and shape of the absorption 'peak' depend on the molecule.

C═O bonds: In aldehydes, ketones, carboxylic acids, esters – sharp deep absorption at 1680–1750 cm^{-1}.

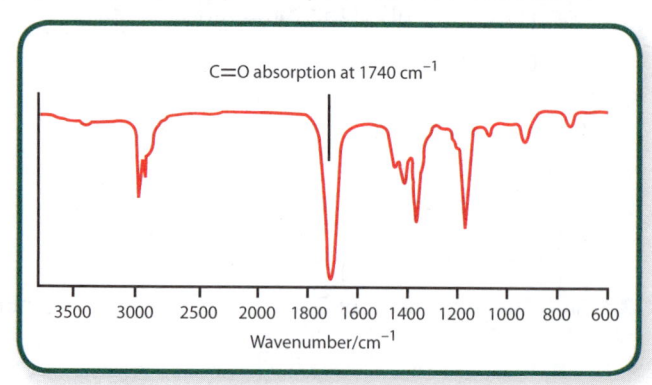

C═O absorption at 1740 cm^{-1}

O—H bonds: Found in alcohols, phenols, carboxylic acids.

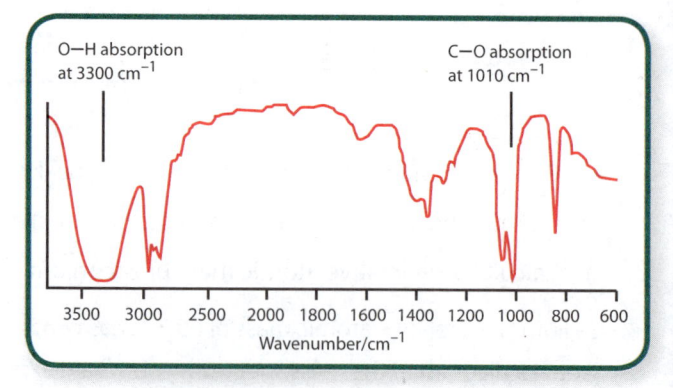

O—H absorption at 3300 cm^{-1} ; C—O absorption at 1010 cm^{-1}

Alcohols and phenols: Show a broad and deep absorption at 3200–3500 cm^{-1}.

Note: If the alcohol is not H bonded (in the gas phase) the peak is much sharper and has a higher wavenumber of 3600–3640 cm^{-1}.

Carboxylic acids: Have a peak of similar shape to that of alcohols but with a lower wavenumber of 2500–3200 cm^{-1}. This overlaps with any C—H bond absorptions.

The spectrum of a carboxylic acid will also show the C═O peak at the low end of the usual range at approximately 1680–1725 cm^{-1}.

The C—O bond (which is also present where there is an —OH group) gives a characteristic absorption at 1000–1300 cm^{-1} but this can be masked as it is in the fingerprint region, e.g. propanoic acid.

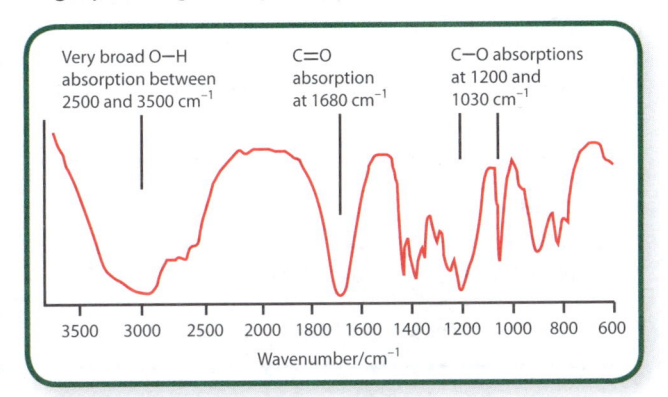

Esters: Show the characteristic C═O and also the more difficult to spot C—O, e.g. ethyl ethanoate.

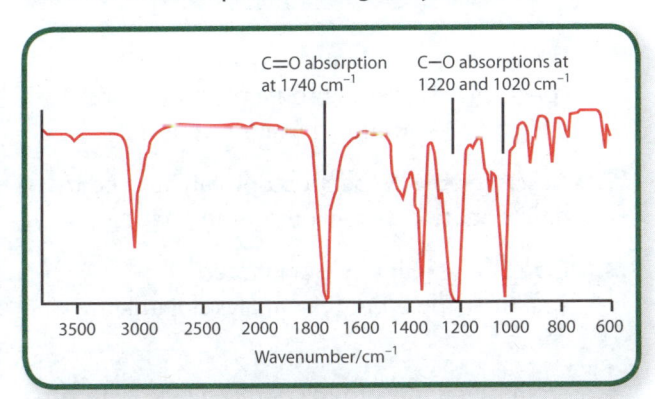

Uses of Infrared Spectroscopy

An infrared spectrum can be used to check the product of an organic reaction.

Example
A secondary alcohol was refluxed with acidified potassium dichromate. The IR spectrum of the product is shown below. What evidence is there that all the alcohol was oxidised?

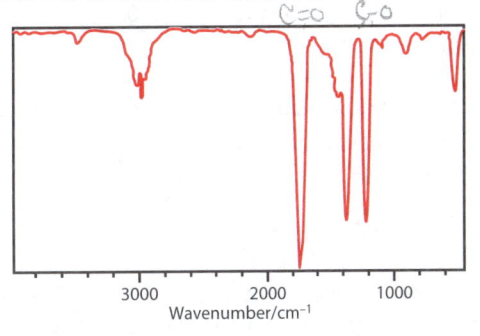

There is no peak at 3200–3500 cm^{-1} indicating no alcohol is present. There is a peak at 1680–1750 cm^{-1} indicating the presence of a C═O group.

Don't forget

…that the absence of a peak may be just as important as the presence of a peak in justifying an answer.

…when describing a peak, to give the range where it may be found rather than a single wavenumber.

Infrared Absorption and Global Warming

UV and visible radiation from the Sun pass through the gases of the atmosphere and reach the Earth. This warms the Earth, which then radiates infrared radiation out towards space.

Some molecules in the atmosphere absorb the infrared energy that is radiated from the Earth's surface. This causes their bonds to vibrate more. These are known as greenhouse gases.

DAY 7

Eventually these molecules lose the extra energy by radiating infrared in all directions, or by colliding with and passing energy on to other molecules.

The collisions with other molecules causes an increase in kinetic energy of the molecules and so an increase in temperature.

The energy received from the Sun and lost by the Earth to space settles to a steady state where the amount received and the amount lost is the same. The position of this equilibrium determines the average temperature of the planet.

Water vapour and carbon dioxide are naturally present in the Earth's atmosphere and have bonds that absorb infrared radiation.

$$\text{water O—H} \quad \text{carbon dioxide C=O}$$

The percentage of carbon dioxide in the atmosphere has increased since the industrial revolution. In the three decades from 1980 to 2010 it rose from around 330 to 3900 parts per million.

This has upset the previous energy equilibrium and is causing more heat to be retained by the Earth's atmosphere and less to be radiated to space.

Most scientists agree that the rise is due to human activity, particularly energy production.

Other gases in the atmosphere have increased due to human activity and are also greenhouse gases. These include methane, CFCs and nitrogen oxide. The percentage increase of these gases is much less than that of carbon dioxide but the amount of energy absorbed per molecule is greater (higher greenhouse factor).

Controlling Carbon Dioxide Levels

Large amounts of carbon dioxide have been added to the atmosphere as a result of the combustion of fossil fuels. Use of renewable sources of energy reduces the amount of carbon released.

This has prompted many governments to adopt policies that increase the use of renewable energy and decrease the use of fossil fuels.

Calculating Parts Per Million

Gases with very low concentrations, such as trace gases in the atmosphere, are measured in parts per million.

$$1 \text{ ppm} = 0.0001\% \quad 1\% = 10\,000 \text{ ppm}$$

To convert from parts per million to per cent, divide by 10 000.

To convert from per cent to parts per million multiply by 10 000.

Example
The concentration of CO_2 in the atmosphere is 0.04%. What is this is in parts per million?

$$0.04 \times 10\,000 = 400 \text{ ppm}$$

QUICK TEST

1. What difference would be seen in the IR spectrum of a sample containing a low and a high concentration of water vapour?

2. Describe how IR spectroscopy can be used to differentiate between hexane and heptane.

3. Give one practical use of infrared spectroscopy other than analysis of unknown compounds.

4. What functional group is indicated by a strong absorption in the range 1700–1750 cm^{-1}?

5. How could you identify an alcohol from its IR spectrum?

6. What is 25 ppm expressed as a percentage?

114

PRACTICE QUESTIONS

1. An alcohol was refluxed with acidified potassium dichromate. The infrared spectrum of the product showed a peak at 3200–3500 cm^{-1} but no peak at 1700–1750 cm^{-1}. The alcohol was **[1 mark]**

   A primary   B secondary
   C tertiary   D either primary or secondary.

2. The fingerprint region of an infrared spectrum lies between **[1 mark]**

   A 1000 and 1500 cm^{-1} B 500 and 1000 cm^{-1}
   C 800 and 1600 cm^{-1} D 500 and 1500 cm^{-1}.

3. Infrared spectroscopy can be used to measure the alcohol content of breath. Breath contains a mixture of molecules including nitrogen, oxygen and water vapour. The alcohol in alcoholic drinks is ethanol.
   a) What characteristic IR absorptions would be seen from a sample of pure ethanol? **[3 marks]**
   b) How could the spectrum be used to determine the concentration of ethanol? **[1 mark]**
   c) Which of the other components of breath might interfere with the measurement of ethanol and why? **[2 marks]**
   d) Suggest how ethanol could be identified despite this interference. **[2 marks]**

4. Scientists have used infrared spectroscopy to measure the concentration of carbon dioxide in the atmosphere. In December 2014 the level was measured as 389.62 ppm. When carbon dioxide was measured as a percentage in December 2015 the value was 0.040162%.
   a) What feature of carbon dioxide makes it suitable for analysis by infrared spectroscopy? **[1 mark]**
   b) What happens to carbon dioxide molecules when they absorb infrared radiation? **[1 mark]**
   c) Explain why the ability of carbon dioxide to absorb infrared energy means that it is a greenhouse gas. **[2 marks]**
   d) What is the percentage increase in carbon dioxide in the atmosphere between December 2014 and December 2015? **[1 mark]**

5. A student prepared a sample of propanoic acid by oxidising propan-1-ol. A spectrum of the product is shown below.

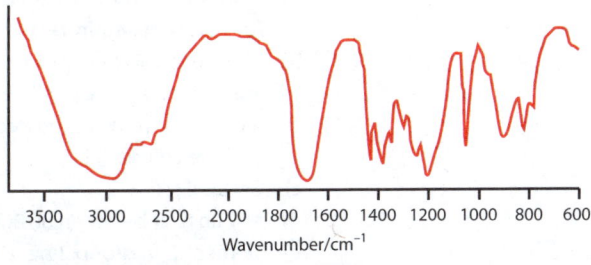

   a) What evidence from the spectrum suggests that the conversion was successful? **[2 marks]**
   b) One possible side product of the reaction is propanal. What characteristic peak would you expect to see in the infrared spectrum of propanal. **[1 mark]**
   c) Explain why it is difficult to tell from the spectrum whether 100% of the alcohol was converted. **[2 marks]**

115

Answers

Day 1
Atomic Structure
QUICK TEST (Page 6)
1. The mass of the atom relative to $\frac{1}{12}$ mass of an atom of carbon-12.
2. 8
3. 18
4. 1
5. Cathode rays
6. Alpha particles were deflected through 180° when fired at gold foil.

PRACTICE QUESTIONS (Page 6)
1. C [1]
2. D [1]
3. B [1]
4. a) 95 [1]
   b) It contains a **different number of neutrons** from some other atoms of americium [1].
   c) ^{241}Am has one more proton [1]; ^{242}Pu has 2 more neutrons [1].
   d) All atoms of Am must have the same atomic number / number of protons [1]. The decay product has two fewer protons / different atomic number [1].
5. a) 7 protons [1] 8 neutrons [1]
   b) There is no difference [1].
   c) They have a different physical properties, e.g. density [1].
   d) It is very small because it has not affected the relative atomic mass [1].
6. Billiard ball model – no sub-atomic structure; atoms of different elements are different.
   Current model – sub-atomic particles; atoms of different elements are different.
   [4 marks: 3 marks for point made above; 1 mark for any other reasonable comparison]

Representing Chemical Reactions
QUICK TEST (Page 10)
1. 8
2. K_2SO_4
3. $2H_3PO_4 + 3Mg \rightarrow Mg_3(PO_4)_2 + 3H_2$
4. $Mg(NO_3)_2 + Na_2CO_3 \rightarrow MgCO_3 + 2NaNO_3$
5. $Pb^{2+}(aq) + 2I^-(aq) \rightarrow PbI_2(s)$
6. $Zn(s) \rightarrow Zn^{2+}(aq) + 2e^-$
   $2H^+(aq) + 2e^- \rightarrow H_2(g)$

PRACTICE QUESTIONS (Page 11)
1. D [1]
2. A [1]
3. a) $NaHCO_3$ [1]
   b) $2NaHCO_3 + H_2SO_4 \rightarrow Na_2SO_4 + 2CO_2 + 2H_2O$
   [2 marks: 1 mark for correct formula Na_2SO_4; 1 mark for balanced equation]
   c) $HCO_3^-(aq) + H^+(aq) \rightarrow CO_2(g) + H_2O(l)$
   [3 marks: 1 mark for correct formula; 1 mark for balanced equation; 1 mark for state symbols]
4. a) TiO_2 [1]
   b) Titanium(IV) chloride
   [2 marks: 1 mark for correct words: 1 mark for correct numeral]
   c) $TiO_2 + 2Cl_2 + 2C \rightarrow TiCl_4 + 2CO$
   [2 marks: 1 mark for correct formula: 1 mark for correctly balanced equation]
   d) Mg^{2+} [1]
   e) $Mg \rightarrow Mg^{2+} + 2e^-$ [1]
   f) Titanium(IV) chloride + hydrogen → titanium(III) chloride + hydrogen chloride
   [2 marks: 1 mark for hydrogen chloride, not hydrochloric acid; 1 mark for totally correct equation]

Electron Configuration
QUICK TEST (Page 14)
1. 3
2. 3
3. [diagram of p-orbital along y-axis with x, y, z axes]
4. [↑↓] [↑↓] [↑] [↑] [↑]
5. $1s^2\ 2s^2\ 2p^6\ 3s^2\ 3p^5$
6. Vanadium

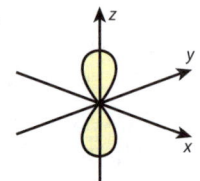

PRACTICE QUESTIONS (Page 15)
1. A [1]
2. B [1]
3. a) $C^+(g) \rightarrow C^{2+}(g) + e^-$
   [2 marks: 1 mark for correct state symbols; 1 mark for correct remainder]
   b) Each successive electron is being removed from an ion with one more positive charge [1]. More energy is required to remove the negative electron from a bigger positive charge [1].
   c) Group 4 [1]
   d) Any answer between 80 000 and 90 000 kJ mol^{-1} [1].
   e) Element is in group 2 because of the large increase in ionisation energy at 3rd ionisation [1]. First ionisation energy is lower than the first ionisation energy of magnesium therefore is in a higher period than magnesium [1] because first ionisation energy decreases down a group [1]. Element is calcium [1].

4. a) First ionisation energy increases across a period [1] because there is one extra proton in the nucleus with similar shielding (smaller atomic radius) [1].
   b) First ionisation energy increases from Li to Be but drops at B [1]. This is because an electron is being removed from a p sub-shell, which is higher in energy than an s sub-shell [1] so requires less energy than removing an electron from a full s sub-shell [1]. This shows that the second shell of electrons is divided into two subshells [1].

Calculations 1
QUICK TEST (Page 18)
1. 219.1
2. 5.60 g
3. 6.25×10^{-3}
4. 0.1 moles
5. 3 : 2
6. 4 g dm^{-3}

PRACTICE QUESTIONS (Page 19)
1. B [1]
2. A [1]
3. a) To make sure that all the water had been driven off [1].
   b) The crucible had reached a constant mass / there was no loss of mass between the 3rd and 4th weighing [1] so all the water had been driven off [1].
   c) Initial mass of sample = 23.78 − 20.05 = 3.73 g
      Mass of CaCl$_2$ after heating = 21.94 − 20.05 = 1.89 g [1]
      M$_r$ of CaCl$_2$ = 111.1 [1]
      Moles of CaCl$_2$ after heating = 1.89/111.1 = 0.0170 [1]
      Mass of water lost = 3.73 − 1.89 = 1.84 g
      M$_r$ of H$_2$O = 18   moles H$_2$O lost = 1.84/18 = 0.102 [1]
      Mole ratio CaCl$_2$: H$_2$O = 0.102/0.017 = 6.01   X = 6 [1]
4. a) M$_r$ of Al$_2$O$_3$ = 102   A$_r$ Al = 27 [1]
      $1 \times 10^6/27 \times 0.5 \times 102 \times 10^6 = 1.89$ tonnes Al$_2$O$_3$ [1]
   b) Moles O$_2$ produced = $1 \times 10^6/27 \times 3/4$
      = 2.78×10^4 moles [1]
      Moles carbon consumed = 2.78×10^4 [1]
      Mass carbon consumed = 2.78×10^4 moles $\times$ 12
      = 3.33×10^5 g = 333 kg
      ∴ 300 kg of carbon are lost from the electrode [1].
   c) More frequently since more moles of carbon are used up per mole of oxygen / mole ratio is 1 : 1 not 2 : 1 [1].
5. Reaction equation NaHCO$_3$ + HCl → NaCl + CO$_2$ + H$_2$O [1]
   $20.05 \times 10^{-3} \times 0.1 = 2.01 \times 10^{-3}$ moles of HCl [1]
   Mole ratio 1 : 1   0.00201 moles NaHCO$_3$ in 25 cm^3 = 0.02 mol in 250 cm^3 [1]
   M$_r$ NaHCO$_3$ = 84   Mass of NaHCO$_3$ in 250 cm^3 84 $\times$ 0.0201
   = 1.69 g
   % NaHCO$_3$ in 2.0 g sample = 1.69/2.0 $\times$ 100 = 84% so the claim is false [1].

Day 2
Ionic Bonding and Structure
QUICK TEST (Page 22)
1. [Mg]$^{2+}$ [:Ö:]$^{2-}$
2. Al^{3+}
3. Li$^+$
4. MgCl$_2$
5. Na$^+$, Mg^{2+} or Al^{3+}
6. The cations move towards the cathode and are discharged / reduced. The anions move towards the anode and are discharged / oxidised.

PRACTICE QUESTIONS (Page 23)
1. D [1]
2. A [1]
3. B [1]
4. a) [diagram of Ca^{2+} ion and 2 Cl$^-$ ions]
   [4 marks: 1 mark for 2 Cl; 1 mark for each correct arrangement of electrons in the ions; 1 mark for correct charges]
   b) Ionic substances have very high melting points [1] because of strong electrostatic attraction between ions [1].
   c) Giant structure [1] diagram shows each ion surrounded by 8 ions of the opposite charge [1] in a 3D structure [1].
   d) Any three from: Potassium has a single charge while calcium has a double charge; potassium ions have fewer protons in the nucleus than calcium ions; calcium has a smaller ionic radius than potassium; calcium has a higher charge density than potassium; increased charge density increases the strength of ionic bonding; calcium chloride has stronger ionic bonding than potassium chloride; calcium chloride has a higher melting point than potassium chloride because more energy is needed to break the ionic bonding / electrostatic attraction.
   [3 marks: 1 mark for each point made]
   e) Sodium has metallic bonding [1] and a giant structure [1] with delocalised electrons that can move to carry an electric current [1]. Sodium chloride has ionic bonding [1] with giant structure [1] and as a solid there are no charged particles that can move to carry the charge [1].
5. a) PbCrO$_4$ [1]
   b) Lead nitrate / ethanoate solution [1]; sodium / potassium / ammonium chromate solution [1]
      Pb^{2+}(aq) + CrO$_4^{2-}$(aq) → PbCrO(s) [1] [State symbols essential; allow error carried forward on formula]
   c) Mix together the lead solution and chromate solution. Filter [1] then wash and dry [1] the precipitate.

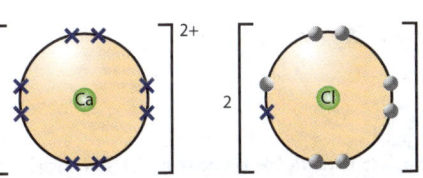

117

Covalent Bonding and Structure
QUICK TEST (Page 26)
1. A shared pair of electrons between nuclei
2. H:C:C:H (H₂C=CH₂ dot-cross diagram with H atoms)
3. Ammonium ion, any complex ion, aluminium chloride, carbon monoxide
4. It has weak intermolecular bonds between molecules that can be broken by the energy available at room temperature.
5. $^{\delta+}C—Br^{\delta-}$
6. Simple covalent/moleculer structures form small discrete molecules with a set ratio and number of atoms. There are weak intermolecular forces of attraction between the molecules that hold the particles together in a solid structure. Giant covalent structures contain very large numbers of atoms held together by strong covalent bonds.

PRACTICE QUESTIONS (Page 27)
1. D [1]
2. A [1]
3. a) O::C::O (dot-cross diagram of CO₂)
   [2 marks: 1 mark for double bonds; 1 mark for lone pairs]
   b) A sigma bond [1] is a shared pair of electrons that are in a direct line between the two nuclei of the sharing atoms [1]. Pi bonds [1] are a shared pair of electrons formed from the overlap of p orbitals above and below a direct line between the nuclei [1].
   c) Both electrons in the bond come from the same atom [1].
   d) N≡N→O
   [2 marks: 1 mark for triple bond; 1 mark for dative covalent bond]
4. a) The two atoms sharing electrons have different electronegativities [1].
   b) Cl—Cl, Br—H, F—H [1]
5. a) Br:Br (dot-cross diagram) [1] Bromine has 7 electrons in its outer shell [1]. By sharing an electron with another bromine atom both gain a full outer shell [1].
   b) Bromine is a simple covalent molecule [1] with weak intermolecular forces between molecules [1] that are easily overcome at a low temperature [1].
   c) $^{\delta+}Br—F^{\delta-}$ [1]
   d) Bromine is more soluble in hexane than in water so the strength of the bonding between solvent molecules is similar to the strength of the bonding between simple covalent molecules and the solvent [1]. The strength of the bonds between water and bromine is not sufficient to overcome the hydrogen bonding between water molecules [1].

Metallic Bonding and Structure/Titration Techniques
QUICK TEST (Page 30)
1. (diagram of Mg²⁺ ions surrounded by e⁻ delocalised electrons)
2. Electrostatic attraction between positively charged metal ions and the negative sea of delocalised electrons.
3. The charge density of Na⁺ is higher than that of K⁺ ions in the metal lattice because Na⁺ has the smallest radius. This means a stronger attraction between the ions and the delocalised electrons in Na than in K and so more energy is needed to break the attraction and melt the metal.
4. The delocalised electrons can move through the metal lattice when a potential difference is applied across it.
5. 2. Record the final digit as 5 or 0 (0.05 or 0.00)
6. Methyl orange

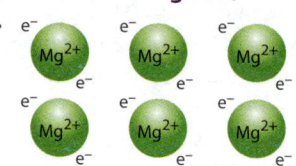

PRACTICE QUESTIONS (Page 31)
1. B [1]
2. D [1]
3. a) Metallic bonding / lattice of metal ions in a sea of delocalised electrons [1]; giant structure [1].
   b) The delocalised electrons [1] are able to move when a potential difference is placed across them [1].
   c) **Any one from:**
   Magnesium would be weaker / softer.
   Magnesium would have a lower melting point.
   Any one from:
   Only two delocalised electrons compared to three for aluminium / lower charge on the ion for magnesium.
   Larger ionic radius for magnesium than aluminium.
   [2 marks: 1 mark for each point made]
4. a) To remove any drops of water or other solution in the burette [1].
   b) To make any colour change easier to see [1].
   c) To avoid any parallax error [1].
   d) To ensure that the analyte comes into rapid contact with the titrant [1].
   e) To remove the last (calibrated) drop from the pipette [1].
   f) To get an accurate volume at the endpoint [1].
   g) To improve the reliability of the results [1].
5. a) **Any four from:** Grind the candy to a powder in a pestle and mortar.
   Weigh a sample by difference.
   Stir the sample in water until dissolved and transparent.
   Pour the sample into a volumetric flask.
   Rinse the beaker into the flask.
   Make up to the mark and mix.
   [4 marks: 1 mark for each point made]

b) Any strong alkali, e.g. KOH or NaOH [1]
   Phenolphthalein [1]
c) Good accurate method for determining acid concentration.
   Not good because other acids may be present.
   Colours in the sweet may interfere with the indicator.
   [1 mark for any sensible comment]

Structures of Carbon/Shapes of Molecules
QUICK TEST (Page 34)
1. Diamond contains no charged particles that are able to move.
2. Delocalised electrons between the layers of carbon atoms are able to move when a potential difference is applied.
3. Buckminsterfullerene
4. 180° – two double bonds around the central carbon
5. Octahedral, 6 bonding pairs.
6. Trigonal pyramidal. One lone pair and three bonding pairs around central phosphorus.

PRACTICE QUESTIONS (Page 35)
1. A [1]
2. D [1]
3. a) Three strong covalent bonds [1] require a lot of energy to break [1].
   b) It dissolved in a solvent [1]. It has a defined number of atoms in its formula [1].
   c) Diamond has 4 single covalent bonds on each carbon [1]; buckminsterfullerene has some carbon–carbon double bonds [1].
   d) Delocalised electrons throughout the whole structure [1] are able to move and carry the charge when a potential difference is applied [1].
4. a) 109.5° [1]
   b) 4 regions of electron density [1] around the carbon repel and move as far apart as possible [1].
   c) Tetrahedral [1]
   d) 107° [1]
   e) The N atom has a lone pair but the C atom does not [1]. The lone pair is more repelling than a bonding pair [1].
5. a)

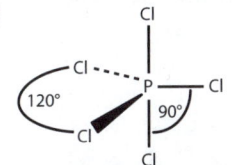

   Trigonal bipyramidal
   [3 marks: 1 mark for each angle (maximum of 2); 1 mark for name of the shape of the molecule]
   b) Tetrahedral [1]. There are 4 groups of electron density around the phosphorus [1].

Day 3
Intermolecular Forces
QUICK TEST (Page 38)
1. The molecules have more electrons so there are stronger instantaneous dipole–induced dipole forces.
2. 2,2-dimethylpropane; more branching means weaker intermolecular bonds
3. C—F, C—N, C—O, I—F, S—H
4. CCl_4, CO_2
5.

6. There is no hydrogen covalently bonded to F, O or N.

PRACTICE QUESTIONS (Page 39)
1. D [1]
2. B [1]
3. A [1]
4. Molecules with larger numbers of electrons/M_r have stronger instantaneous dipole–induced dipole forces [1]. This gives them higher boiling points [1] as the energy required to break the forces is greater [1].
5. a) Ethanol – hydrogen bonding [1]; ethanal – permanent dipole–permanent dipole forces [1]; ethanoic acid – hydrogen bonding [1].
   b) Permanent dipole–permanent dipole forces are weaker than hydrogen bonding [1] so ethanal will have a lower boiling point and vaporise at a lower temperature than the H bonded ethanol [1].
   c) Both molecules can hydrogen bond with the water molecules [1]. This provides the energy needed to disrupt the hydrogen bonding in water [1].
6. All molecules in ice have two H bonds forming an open network; in water fewer H bonds per molecule allows the spaces in the network to be filled in. [4 marks: 1 mark for diagram showing two hydrogen bonds per water molecule; 1 mark for spaces between bonds; 1 mark for each point made (maximum of 2)]

Energetics
QUICK TEST (Page 42)
1.

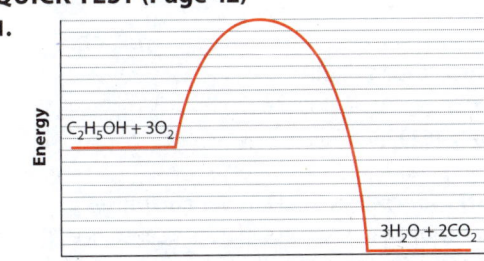

   [Words can also be used for reactants and products]
2. Standard enthalpy of combustion of methane
3. Two moles of butanol are formed.

119

4. 0.5H$_2$SO$_4$(aq) + NaOH(aq) → 0.5Na$_2$SO$_4$(aq) + H$_2$O(aq)
5. 1254 J/1.25 kJ
6. It will be too low.

PRACTICE QUESTIONS (Page 43)
1. D [1]
2. D [1]
3. a) Starting and maximum temperature of the water [1]; mass of water [1]; starting and final mass of spirit burner [1].
   b) −3400 kJmol^{-1}
   [2 marks: 1 mark for sign and units; 1 mark for correct value]
   c) Lower [1] than true value due to heat loss to the surroundings/incomplete combustion/heating of the apparatus/not taken under standard conditions [1].
   d) Use a draft shield/ensure adequate supply of oxygen [1].
   e) Incomplete combustion [1] reduces the value for enthalpy of combustion [1].
4. Energy change = 50 × 4.18 × 6.8 = −1.421 kJ [1]
   H$_2$SO$_4$ + 2KOH → 2H$_2$O + K$_2$SO$_4$ or mole ratio acid : alkali = 1 : 2 [1]
   Moles water made = $\frac{25 \times 10^{-3}}{0.025}$ [1]
   Enthalpy of neutralisation = $\frac{-1.421}{0.025}$ = −56.84 kJ mol^{-1}
   [2 marks: 1 mark for correct value; 1 mark for correct units]

Hess's Law
QUICK TEST (Page 46)
1. 0.5O$_2$(g) + H$_2$(g) → H$_2$O(l)
2. Because the balanced formula equation for both is the same: C(s) + O$_2$(g) → CO$_2$(g)
3. C$_3$H$_7$OH(l) + 4.5O$_2$(g) → 3CO$_2$(g) + 4H$_2$O(l)
   ↑ 3C(s) + 4H$_2$(g) + 5O$_2$(g)
4. C—H C—C C—O O—H C═O
5. O$_2$(g) is already the product in the definition.
6. 1.25%

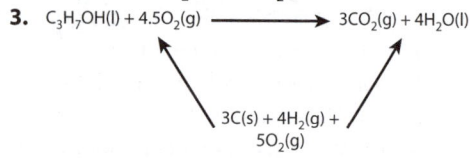

PRACTICE QUESTIONS (Page 47)
1. C [1]
2. D [1]
3. a) ΔH$^\ominus$$_r$ = −ΔH$^\ominus$$_f$(CaCO$_3$) + ΔH$^\ominus$$_f$(CaO) + −ΔH$^\ominus$$_f$(CO$_2$)
   [All correct for 1 mark]
   b) −(−1207) + (−635) + (−394)
   = 1207 − 1029
   = +178 kJ mol^{-1}
   [3 marks for correct final answer and units]
4. a) The enthalpy change when 1 mole of a substance [1] is formed from its elements in their standard state [1] under standard conditions/at 100 kPa/1 atm and 298 K [1].

   b) 3C(s) + 3H$_2$(g) + 0.5O$_2$(g) → CH$_3$COCH$_3$(l)
      ↓ 4O$_2$(g)   ↓ 4O$_2$(g)
      3CO$_2$(g) + 3H$_2$O(l)

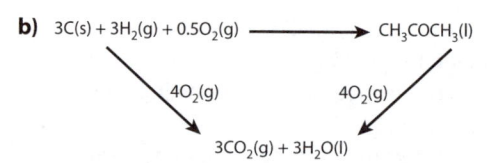

   ΔH$^\ominus$$_f$ (propanone) = 3(−394) + 3(−286) − (−187)
   = −1853 kJ mol^{-1}
   [4 marks: 1 mark for correct enthalpy cycle; 1 mark for correct state symbols; 1 mark for final answer; 1 mark for correct units]
5. a) C$_6$H$_{14}$
   [1 mark for correct formula for hexane]
   14 × C—H = 14 × 413 = 5782
   5 × C—C = 5 × 347 = 1735
   9.5 × O═O = 9.5 × 497 = 4722
   Total bonds broken = 12 239

   12 × C═O = 12 × 805 = 9660
   14 × O—H = 14 × 463 = 6482
   Total bonds formed = −16 142

   Enthalpy change = 12 239 − 16 142 = −3903 kJmol^{-1}
   [2 marks for correct calculations leading to correct final answer]
   b) The values are for average bond enthalpies not for these specific molecules [1]. The values are taken from the gaseous state; hexane is a liquid [1].

Kinetics
QUICK TEST (Page 50)
1. Temperature, concentration, pressure for gases, surface area for solids, presence of a catalyst.
2. The concentration of reactants decreases.
3. If the activation energy is low the energy of the surroundings is sufficient to start the reaction; if the activation is high, energy must be put in.
4. Homogeneous catalysts are in the same physical state/phase as the reactants while heterogeneous catalysts are in a different state/phase.
5. Measure the volume of gas produced over time.
6. [Graph showing Concentration vs Time with two curves]

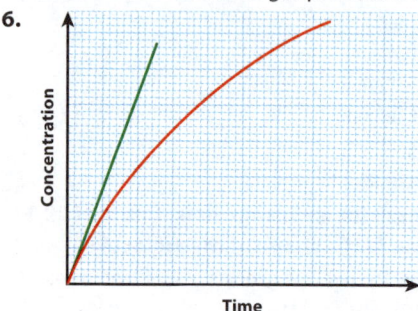

PRACTICE QUESTIONS (Page 51)
1. B [1]
2. C [1]

120

3.

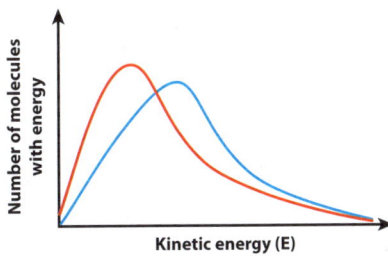

a) The peak of the curve is to the right [1] and lower [1]; the line remains above the original once past the peak [1].
b) The activation energy is to the right of the peak/shown as a line on the graph [1]; at the higher temperature more particles have greater than or equal to the activation energy for the reaction [1].
c) The shape of the curve would be the same [1] but all number/frequency values would be higher [1].

4. a) A catalyst speeds up the rate of the reaction [1] but is chemically unchanged by the reaction [1].
b) Homogeneous catalyst [1]
c)

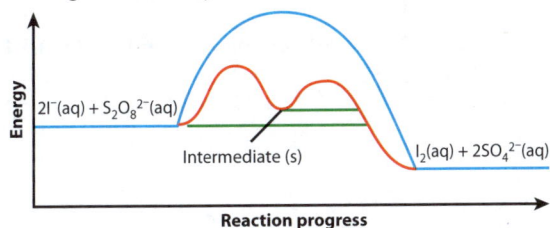

Reactants higher than products [1]; higher line joining reactants and products labelled 'uncatalysed reaction' [1]; lower line of any shape labelled 'catalysed reaction' [1]; intermediate labelled at lower energy than two peaks either side [1].

d) Either
Iodine is coloured [1]; plot absorbance against time for the production of iodine [1]. Take a tangent to the curve at time = 0 [1]. Calculate the gradient [1].
Or
Add a known quantity of thiosulfate [1] and some starch indicator [1]. Measure the time taken for the mixture to turn blue-black [1]. The rate is inversely proportional to the time [1].
[Maximum of 4 marks]

5. a) To increase the surface area [1] without using large quantities of expensive metals [1].
b) The rate of reaction is very slow when the temperature is low [1].
c) Heterogeneous catalyst [1]
d) Reactant particles are adsorbed onto a solid catalyst surface [1]. The bonds between the reactant atoms weaken and break [1]. New bonds form between the product atoms [1]. The products desorb and diffuse away from the catalyst surface allowing new reactants to take their place [1].

Day 4

Calculations 2

QUICK TEST (Page 54)

1. CH_2
2. $C_4H_8O_2$
3. 94.6%
4. 60.9%
5. 2.4 dm^3
6. 2.0

PRACTICE QUESTIONS (Page 55)

1. C [1]
2. D [1]
3. C [1]
4. a) 1.64 g [1]
   b) 2.5 ÷ 24.3 = 0.103 [1]   1.64 ÷ 16 = 0.103 [1]   MgO [1]
5. a) Route 1 [1]
   b) Theoretical yield: 200 ÷ 60 = 3.3 moles [1] 88 × 3.3 = 290.4 g [1] (190 ÷ 290.4) × 100 = 65.4% [1]
6. a) 0.6 ÷ 24 [1] = 0.025 moles [1]
   b) Volume ratio = mole ratio hydrocarbon to carbon dioxide 1800 ÷ 600 [1] = 3 moles carbon dioxide per mole hydrocarbon [1]
   c) Mole ratio H_2O to hydrocarbon = 3 : 1 [1]. Each mole of water comes from 2 hydrogen atoms so hydrocarbon contained 6 H atoms [1]. Formula of propane is C_3H_8 [1] so the hydrocarbon is not propane [1].
7. a) M_r = 65 [1] moles = 2 [1]
   b) 72 dm^3 [1]
   c) p = nRT ÷ V = (2 × 8.31 × 293) ÷ 0.065
   [2 marks: 1 mark for correctly rearranged equation; 1 mark for volume converted to m^3]
   = 74917.8 Pa /74.9 kPa
   [2 marks: 1 mark for correct number; 1 mark for correct units]
   [Maximum of 4 marks]

Chemical Equilibria

QUICK TEST (Page 58)

1. For a system at equilibrium, the concentration of reactants and products will change to oppose any change made to the system.
2. Equilibrium moves to the right.
3. Equilibrium moves to the left.
4. No change
5. No change
6. For economic reasons such as costs of achieving high temperatures and pressures.

PRACTICE QUESTIONS (Page 59)

1. C [1]
2. D [1]
3. a) It would become more red [1].
   b) Add alkali / add $Cr_2O_7^{2-}$(aq) [1].

121

4. **a)** It is an equilibrium reaction [1] so the backward reaction will reform water from hydrogen [1].
   b) Low pressure [1] – there are more moles of gas in the products than in reactants [1].
   c) It speeds up the rate at which equilibrium is achieved [1] without additional energy costs [1].
   d) The carbon monoxide reaction is exothermic [1] so low temperatures give a higher equilibrium yield of hydrogen [1]. The methane reaction is endothermic so high temperatures give the best yield [1].
   e) The equilibrium is far to the right, on the product side [1].
   f) $K_C = \dfrac{[CO][H_2]^2}{[CH_4][H_2O]}$

   [3 marks: 1 mark for products on top; 1 mark for value of hydrogen concentration squared; 1 mark for completely correct]

Redox
QUICK TEST (Page 62)
1. Reduced
2. Copper
3. Oxidation
4. Chlorate(V)
5. NO_2^-
6. Hydrogen and oxygen

PRACTICE QUESTIONS (Page 63)
1. C [1]
2. C [1]
3. B [1]
4. A [1]
5. **a)** $2Br^- + Cl_2 \rightarrow 2Cl^- + Br_2$ [1] [Ignore state symbols]
   b) Br^- [1]
   c) Oxidation state of silver in $AgNO_3$ = +1   AgBr = +1 [1]
   d) It reduces silver $Ag^+ + e^- \rightarrow Ag$ [1] [Equation not essential]
6. **a)** Titanium(IV) chloride [1]
   b) The same element is oxidised and reduced [1].
   $\qquad\qquad\quad 2TiCl_3 \rightarrow TiCl_2 + TiCl_4$
   Oxidation state of titanium +3   +2   +4 [1]
   +3 → +4 is oxidation   +3 → +2 is reduction [1]

Nuclear Reactions & Radiation
QUICK TEST (Page 66)
1. Thorium
2. A positron
3. Aluminium foil
4. 2.98×10^{-19} J
5. 16 days
6. Coloured lines on a dark background

PRACTICE QUESTIONS (Page 67)
1. B [1]
2. C [1]
3. B [1]
4. **a)** Lead-206 [1]
   b) The radiation is damaging to living cells [1]; α particles are stopped easily and release their energy so will release the energy into the body [1].
   c) The time taken for one half of the radioactive nuclei to decay [1].
   d) 414 days [1]
   e) β particles are able to penetrate outside the body so can be detected [1].
5. **a)** $^2_1H + ^3_1H \rightarrow ^4_2He + ^1_0n$
   [3 marks: 1 mark for each correct species]
   b) Be or any combination of particles that add up to $2 \times ^4_2He$ [1].
6. **a)** 5.75×10^{-19} J / 5.7×10^{-22} kJ
   [2 marks: 1 mark for correct number; 1 mark for correct units]
   b) 8.60×10^{14} Hz
   [2 marks: 1 mark for correct number; 1 mark for correct units]

Day 5
Periodicity/Group 2 Elements
QUICK TEST (Page 70)
1. Any element: B → Ne, Al → Ar, Ga → Kr, In → Xe, Tl → Rn
2. Its highest energy electron is in a d sub-shell.
3. The electron is being removed from a shell that is further from the nucleus, so of higher energy.
4. S_8 molecules have stronger instantaneous dipole–induced dipole forces than P_4 molecules because they have more electrons/have larger molecules/have greater surface area contact/have great molecular mass.
5. pH increases down the group.
6. $2Mg(NO_3)_2(s) \rightarrow 2MgO(s) + 4NO_2(g) + O_2(g)$

PRACTICE QUESTIONS (Page 71)
1. C [1]
2. D [1]
3. a) Group 4 elements have a giant covalent structure [1]. Large amounts of energy [1] are required to break strong covalent bonds [1].
   b) Giant metallic structure [1]; charge of ions increases from 1+ to 3+ [1]; radius of ions decreases [1]; strength of metallic bonding increases [1].
   c) Period 2 metals have one less electron shell so a smaller radius and higher charge density than Group 3 metals [1] giving stronger metallic bonding [1].
   d) Nitrogen, oxygen and fluorine in Period 2 and chlorine in Period 3 all form diatomic molecules [1]. Neon and argon are both monoatomic [1]. Sulfur in Group 6 of Period 3 forms S_8 molecules [1], which have more electrons than any of these and so stronger instantaneous dipole–induced dipole forces [1] requiring more energy to break [1].
4. a)

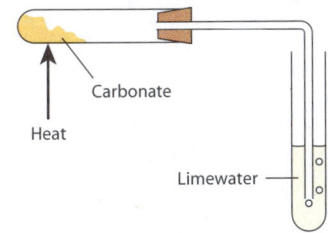

   [2 marks: 1 mark for limewater; 1 mark for remaining set up]
   b) Heat the same quantity of each Group 2 carbonate with the same Bunsen flame [1]; measure the time until the limewater becomes cloudy [1].
   c) The nearer the top of Group 2 the element the shorter the time for the limewater to become cloudy [1].
   d) The charge density of the Group 2 metal ion decreases down the group [1]. The higher charge density at the top of the group distorts the shape of the carbonate ion [1] making it easier to decompose [1].

5. a) $Ba(s) + 2HCl(aq) \rightarrow BaCl_2(aq) + H_2(g)$
   [2 marks: 1 mark for correct products; 1 mark for correct state symbols]
   b) Fizzing/effervescence [1] and the barium would get smaller/disappear [1].
   c) Blue/purple [1]

Group 7: The Halogens
QUICK TEST (Page 74)
1. Purple
2. Increasing number of electrons means increasing strength of instantaneous dipole–induced dipole forces.
3. Colour change from green to brown.
4. $Cl_2 + H_2O \rightarrow HCl + HClO$
5. $2NaBr + 2H_2SO_4 \rightarrow Na_2SO_4 + SO_2 + Br_2 + 2H_2O$
6. Bromide

PRACTICE QUESTIONS (Page 75)
1. A [1]
2. B [1]
3. C [1]
4. B [1]
5. a) $Cl_2(aq) + 2I^-(aq) \rightarrow I_2(aq) + 2Cl^-(aq)$ [1]; chlorine is green, iodine is brown [1].
   b) Chlorine has 3 shells of electrons, bromine has 4 shells of electrons, iodine has 5 shells of electrons [1]. The further the outer shell is from the nucleus the weaker the ability of the atoms to attract electrons to its outer shell [1]. So chlorine is able to take electrons from bromide / oxidise bromide but iodine is not [1].
   c) $8NaI + 5H_2SO_4 \rightarrow 4I_2 + H_2S + 4H_2O + 4Na_2SO_4$ [1] oxidation state of I in NaI is −1 and in I_2 is 0 [1]. $NaCl + H_2SO_4 \rightarrow NaHSO_4 + HCl$ [1] oxidation state of chlorine is −1 in both NaCl and HCl [1].
6. a) $Cl_2 + 2NaOH \rightarrow NaCl + NaClO + H_2O$
   [2 marks: 1 mark for correct formula for NaClO; 1 mark for remainder correct]
   b) The same element is oxidised and reduced in the reaction [1].
   Chlorine is oxidised 0 (Cl_2) → +1 (HClO) [1]
                reduced 0 (Cl_2) → −1 (Cl^-) [1]
   c) HClO [1]
7. The boiling point of the hydrogen halides increases down the group from HCl to HI [1]. HF is anomalous because it is able to hydrogen bond [1]. Hydrogen bonding is stronger than other types of intermolecular forces and requires more energy to overcome [1].

Uses of Group 2 and 7 Elements and Compounds
QUICK TEST (Page 78)
1. To neutralise soil acidity / increase pH of the soil.
2. $2Ca(OH)_2 + 2SO_2 + O_2 \rightarrow 2CaSO_4 + 2H_2O$
3. Opaque to X-rays, insoluble

4. Green
5. A green precipitate forms.
6. Add acidified silver nitrate solution and look for a white precipitate that dissolves in dilute ammonia solution.

PRACTICE QUESTIONS (Page 79)
1. C [1]
2. D [1]
3. a) **Any one from:** Acid rain/natural breakdown of plant material producing hydrogen ions; high crop yields removing basic ions.
      [1 mark for any point made]
   b) $2H^+ + CaCO_3 \rightarrow CO_2 + H_2O + Ca^{2+}$
      [2 marks: 1 mark for correct formula for limestone; 1 mark for remainder correct]
   c) Thermal decomposition and addition of water [1].
   d) It is faster acting / lighter to transport [1].
4. a) To kill microorganisms [1].
   b) **Either** it reacts to form halomethanes [1], which are carcinogenic [1]. **Or** transporting chlorine is a hazard [1] because it is a toxic gas [1].
      [Maximum of 2 marks]
   c) There would be a risk of pathogenic organisms in the water resulting in disease [1]. The cost of water treatment would increase if other sterilising agents were used [1].
5. a) Aluminium ions as there is no flame colour [1] and aluminium hydroxide is a white precipitate that dissolves in excess hydroxide [1].
   b) Not a halide as there is no precipitate with silver nitrate [1]; not a carbonate as there is no reaction with acid [1].
   c) Add barium chloride solution – a white precipitate indicates sulfate ions [1].

Chemistry and the Environment
QUICK TEST (Page 82)
1. They take in carbon dioxide when they grow.
2. Expensive, uses land needed for food production.
3. No polluting waste gases, high energy density, renewable.
4. Collecting and sorting is costly.
5. Releases toxic gases.
6. Starch

PRACTICE QUESTIONS (Page 83)
1. B [1]
2. B [1]
3. a) **Any three from:** Does not release carbon dioxide; does not release polluting gases; does not use up a finite resource; has a higher energy density.
      [3 marks: 1 mark for each point made]
   b) $2H_2(g) + O_2(g) \rightarrow 2H_2O(l)$ [1]
      The product is water, which is also the starting material from which hydrogen is generated [1].
   c) Combustion of hydrocarbons [1] to produce hydrogen would increase the release of polluting gases [1].
   d) Electricity is needed to produce the hydrogen [1]. This is mostly produced by burning fossil fuels, which release carbon dioxide [1].
   e) **Any three from:** Hydrogen is a gas so has a low energy density; it needs to be compressed / stored under pressure; this increases mass/cost/risk; cars have to be designed to use hydrogen; hydrogen fuelling points would have to be built.
      [3 marks: 1 mark for each point made]
4. a) **Any one from:** Build up of non-biodegradable waste (any clear problem related to this); using up of a non-renewable resource.
      [1 mark for any sensible point]
   b) Recycling – expensive/results in low quality product breaking back down to monomers; only suitable for single polymer waste.
      Cracking to produce feedstock – costly and inefficient.
      Incineration – releases toxic gases.
      [2 marks: 1 mark for any one solution; 1 mark for drawback]
   c) Polymers made by or from plants. Do not use crude oil; are generally more biodegradable; absorb carbon dioxide as the plant grows.
      [2 marks: 1 mark for explanation; 1 mark for two advantages]

Day 6
Organic Chemistry
QUICK TEST (Page 86)
1. [cyclohexane with CH₂OH branch structure]
2. Tetrafluoromethane (no numbers needed)
3. [(CH₃)₂CH–Br structure]
4. $C_6H_{12}O$
5. [structure of 3-methylbutanoic acid]
6. Methylpropane
7. 1,1-dichloroethane

PRACTICE QUESTIONS (Page 87)
1. B [1]
2. D [1]
3. a) Haloalkanes/halogenoalkanes [1]
   b) Cl–CF₂–CF₂–Cl [1]
   c) Cl–CClF–CF₂–F [1]
      1,1-dichloro-1,2,2,2-tetrafluoroethane [1]
   d) Trifluoromethane [1]
4. a) Alkene; carboxylic acid
      [2 marks: 1 mark for each functional group]
   b) Butendioic acid
      [2 marks: 1 mark for buten; 1 mark for dioic]
   c) Molecular formula:
      butendioic/fumaric acid is $C_4H_4O_4$ [1]
      pentenoic acid is $C_5H_8O_2$ [1]
      Isomers have the same molecular formula [1].

Alkanes
QUICK TEST (Page 90)
1. $2C_6H_{14} + 19O_2 \rightarrow 12CO_2 + 14H_2O$
2. Alkane vapour is passed over a heated catalyst in the absence of oxygen.
3. To improve the octane number of the fuel.
4. A species with an unpaired electron.
5. UV light provides the energy required to initiate the reaction by homolytic fission of halogens.
6. Many reaction pathways are possible, which give different products.

PRACTICE QUESTIONS (Page 91)
1. A [1]
2. D [1]
3. a) The fractions have different boiling points [1].
   b) D [1] as it has the fewest points of contact between molecules [1] so weaker intermolecular forces [1].
   c) 2,5-dimethylhexane [1]
   d) Reforming [1]
   e) $C_8H_{18} + Br_2 \rightarrow C_8H_{17}Br + HBr$
      [2 marks: 1 mark for correct formula for C_8H_{18}; 1 mark for correct products]
   f) Initiation [1]
   g) The reaction requires bromine radicals [1], which are formed by homolytic fission of bromine in UV light [1].
   h) Radical substitution [1]
4. a) [mechanism showing CH₄ + Cl• → CH₃• + HCl; then CH₃• + Cl–Cl → CH₃Cl + Cl•]
   b) CH₃• + •CH₃ → CH₃–CH₃
      [3 marks: 1 mark for each arrow; 1 mark for correct formula for ethane]
5. a) $C_8H_{18} + 12.5O_2 \rightarrow 8CO_2 + 9H_2O$
      [2 marks: 1 mark for correct formulae; 1 mark for correct balancing]
   b) Any one from: NO causes breathing problems/acid rain/photochemical smog [1]; CO is toxic [1].
   c) Any one from: Platinum Pt; palladium Pd; rhodium Rh [1]
   d) $2CO + 2NO \rightarrow N_2 + 2CO_2$

Haloalkanes
QUICK TEST (Page 94)
1. 1-bromobutane
2. Warm with an aqueous base such as sodium hydroxide, acidify with nitric acid and add silver nitrate solution. A white precipitate, indicates the presence of a chloroalkane.
3. [structure of tertiary haloalkane with Cl]
   Tertiary haloalkane

125

4. Aqueous sodium or potassium hydroxide and reflux
5. An alkene
6. CFCs break down to produce Cl, which acts as a catalyst in ozone depletion.

PRACTICE QUESTIONS (Page 95)

1. D [1]
2. B [1]
3. a) $CH_3CH_2CH_2I + 2NH_3 \rightarrow CH_3CH_2CH_2NH_2 + NH_4I$
   [2 marks: 1 mark for correct reactants; 1 mark for correct products (Maximum of 1 mark if only one NH_3 used as reactant and HI as product)]
   b) Amines [1]
   c) Heat [1] in a sealed tube [1].
   d) Nucleophilic substitution [1]
4. a) [mechanism diagram showing curly arrows from OH⁻ attacking C-Br bond, producing alkene + Br⁻ + H₂O]

   [2 marks: 1 mark for each correct curly arrow (max 2); 1 mark each for correct reactant and product (max 2)]
   b) Hydrogen may be lost from the carbon on either side of the halogen resulting in a different position of the double bond different E/Z isomers can form [1].
   c) A nucleophilic substitution would compete [1] giving an alcohol as a product [1].
5. a) Bromoalkane [1] as it has lower bond enthalpy of C—Br than C—F [1], which makes the reactions easier/faster [1].
   b) 1-bromopentane [1] and KCN [1] in alcoholic solution [1].
   c) Species that can donate a pair of electrons [1] to form a covalent bond to carbon [1] CN⁻ [1].
6. a) Chloroflurocarbons / haloalkanes containing chlorine and fluorine atoms [1].
   b) $O\bullet + O_2 \rightleftharpoons O_3$ [1]
   c) Chloroalkanes/CFCs break down under UV radiation [1].
   R—Cl → R• + Cl• [1] [Radical dots not essential]
   Chlorine radicals **catalyse** the breakdown of ozone [1].
   $O_3 + Cl\bullet \rightarrow ClO + O_2$
   $ClO + O \rightarrow O_2 + Cl\bullet$ [1]

Alkenes

QUICK TEST (Page 98)

1. [zig-zag alkene structure]
2. Pi bonds and sigma bonds
3. An unsaturated molecule contains carbon–carbon double bonds; a saturated compound contains only carbon–carbon single bonds.
4. Shake with bromine water. A colour change from brown to colourless indicates an alkene.
5. Heat 330°C, 6 MPa with steam and H_3PO_4 catalyst.
6. [polystyrene repeat unit structure with C_6H_5 and H groups]

PRACTICE QUESTIONS (Page 99)

1. C [1]
2. D [1]
3. a) A molecule that contains at least one carbon–carbon double bond [1].
   b) Shake with bromine (water) [1]: unsaturated oils give a colour change from brown to colourless [1].
   c) **Either** high temperature and pressure [1] and a nickel catalyst [1] **or** room temperature and pressure [1] and a platinum catalyst [1]. [Maximum of 2 marks]
   d) Electrophilic [1] addition [1]
   e) They have restricted/no rotation around the double bond [1] and two different groups on each carbon on either side of the double bond [1].
4. a) $C_5H_{10} + HCl \rightarrow C_5H_{11}Cl$
   [2 marks: 1 mark for correct reactants and products; 1 mark for correct balancing]
   b) [structures of 2-chloropentane and 1-chloropentane]

   2-chloropentane, 1-chloropentane [4 marks: 1 mark for each correct structure; 1 mark for each name]
   c) 2-chloropentane [1]: The reaction proceeds via a carbocation [1]; the most stable carbocation is formed [1]; 2-chloropentane is formed from a secondary carbocation, which is more stable [1] (than the primary carbocation formed in 1-chloropentane).
5. a) [structure of H and Cl on C=C with H and H] [2]
   b) [polymer structure with CF₂ repeat units]

   [2 marks: 1 mark for correct structure; 1 mark for at least 2 monomer units long]
   c) $CH_2=C(CH_3)-COOH$ [2]
   d) Heat to 200°C at 1500 atm. with a catalyst [1].

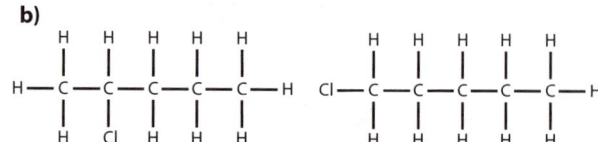

126

Day 7
Alcohols
QUICK TEST (Page 102)
1.

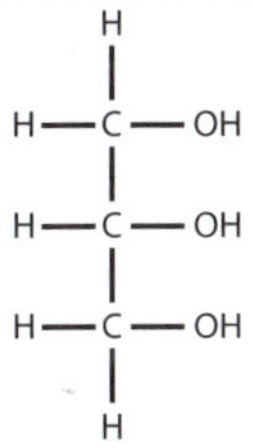

   H—C—OH
   H—C—OH
   H—C—OH
   H

   Both primary and secondary.
2. Concentrated sulfuric acid, reflux / pass alcohol vapour over a heated Al_2O_3 catalyst.
3. No change
4. Oxidation
5. To protonate the –OH group.
6. They can hydrogen bond to water.
7. $CH_3CH_2OH(l) + PCl_5(s) \rightarrow CH_3CH_2Cl(l) + POCl_3(l) + HCl(g)$

PRACTICE QUESTIONS (Page 103)
1. B [1]
2. D [1]
3. A [1]
4. a) and b)

   primary — butan-1-ol

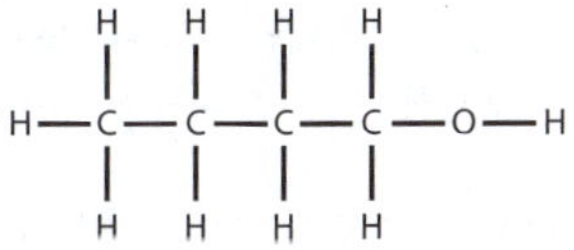

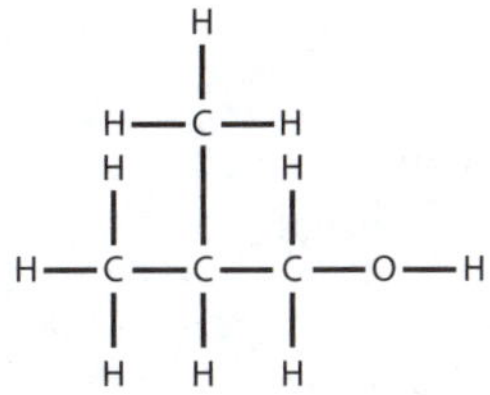

   secondary — butan-2-ol

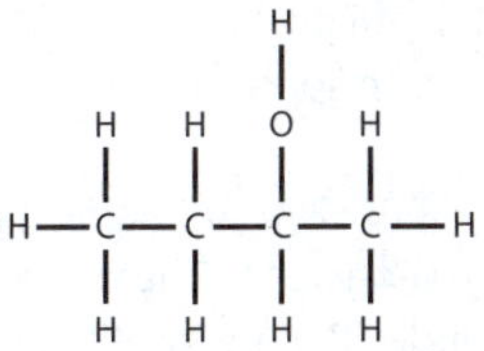

   tertiary — 2-methylpropan-2-ol

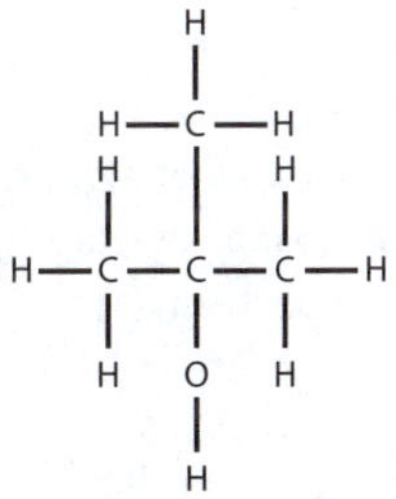

   Also 2-methylpropan-1-ol

   [3 marks for displayed formulae; 3 marks for names]

5.
   c) butan-1-ol → (displayed formula of butanoic acid)

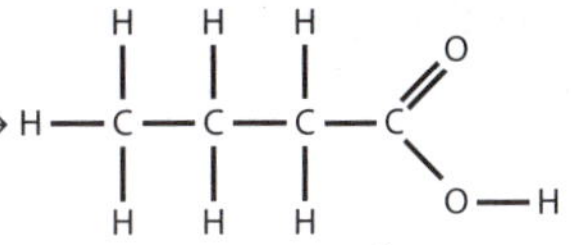

   2-methylbutan-1-ol → (displayed formula of 2-methylbutanoic acid)

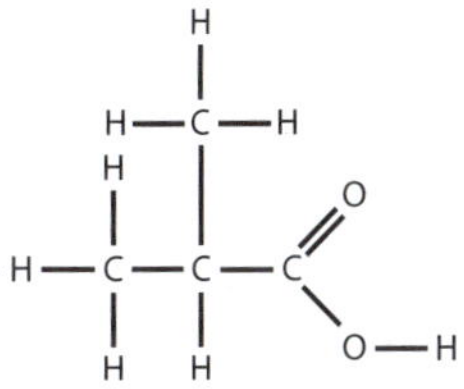

   butan-2-ol → (displayed formula of butanone)

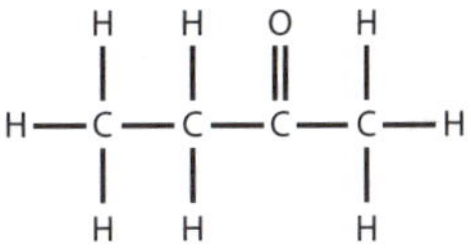

   2-methylpropan-2-ol → no reaction

   [3 marks: 1 mark for each displayed formula]

   a) Concentrated sulfuric acid [1] reflux [1] / pass vapour [1] over a heated Al_2O_3 [1] catalyst. **[Maximum of 2 marks]**

   b) 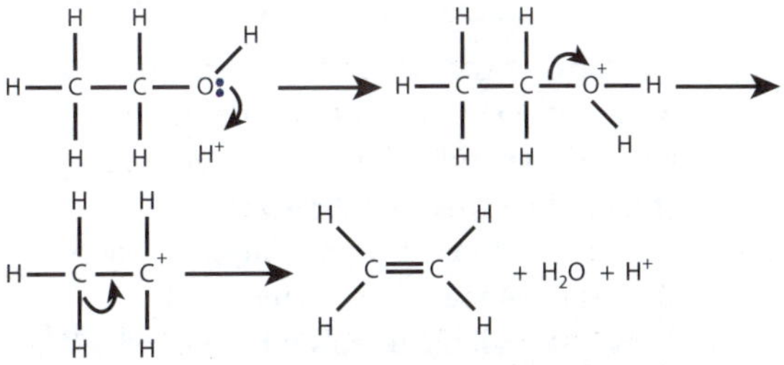 (mechanism diagram)

   [4 marks: 1 mark for each correct curly arrow (max 3); 1 mark for correct reactants and product]

   c) The loss of the second hydrogen can come from the carbon on either side of the alcohol group or E/Z isomers can form. [2 marks for either point made]

6.

Reactant	Reaction conditions	product	Type of reaction
ethanol	$H^+/K_2Cr_2O_7$ reflux	ethanoic acid	oxidation
propanol	pass the vapour over heated Al_2O_3	propene	elimination/dehydration
butan-1-ol	PCl_5 or concentrated HCl	1-chlorobutane	nucleophilic substitution

[1 mark per correct answer (max 6)]

Experimental Techniques
QUICK TEST (Page 106)
1. To heat at boiling point without loss of reactants, products or solvent.
2. At the lowest point.
3. With a solid drying agent.

4. Check the density of the two substances.
5. Recrystallisation
6. If pure it will melt over a narrow range of temperatures.

PRACTICE QUESTIONS (Page 107)
1. C [1]
2. A [1]
3. a)

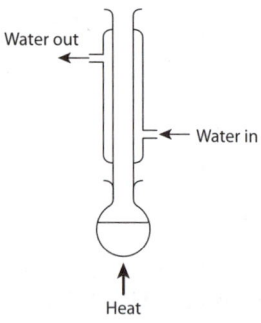

   Diagram to include the following: condenser vertical [1]; water in at bottom, out at top [1]; joint between flask and condenser sealed [1].
   b) The top of the condenser/still head is not sealed [1]; water should be in at bottom and out at top [1].
   c) For better control of heating / to reduce the risk of fire [1 mark for either point made]
4. a) Dissolve the sample in a **minimum** of warm solvent [1]. Decant to remove any solid remains [1].
   Allow the solution to cool and recrystallise [1].
   Filter the crystals under vacuum [1].
   Wash in a minimum **cold** solvent and dry in a cool oven [1].
   b) Pure substances melt over a narrow range of temperatures [1] at the expected melting point [1].

Mass Spectrometry
QUICK TEST (Page 110)
1. [^{79}Br]$^+$
2. Electrons are lost when the sample passes through a high energy electron beam.
3. 5
4. 102 molecular ion; 103 M+1 peak
5. [C$_6$H$_5$]$^+$
6. CH$_3$

PRACTICE QUESTIONS (Page 111)
1. A [1]
2. C [1]
3. a) ^{90}Zr$^+$ [2 marks (Must include positive charge mark)]
   b) Species with different masses gain different acceleration [1] in the electric field [1] and take different times [1] to reach the detector.
   c) 91.3 [1]
4. Peaks at 69 large than peak at 71 [1]; x-axis labelled m/z [1].

5. a) Molecular ion formula contains oxygen [1]; [C$_3$H$_8$O]$^+$ [1]; base peak [CH$_3$O]$^+$ [1].
   b) C$_2$H$_5$• [1]
   c) [1 mark for any structure with formula C$_3$H$_8$O]

   H—C—C—C—O—H [1]

   d) Check against the known spectrum of propan-1-ol [1].

Infrared Spectroscopy
QUICK TEST (Page 114)
1. A high concentration gives a deeper peak in the O—H region.
2. Take an IR spectrum and compare it to standard spectra.
3. Measurement of atmospheric gases, vehicle emissions, alcohol in breath.
4. A carbonyl group
5. A strong absorption in the range 3200–3600 cm^{-1}
6. 0.0025%

PRACTICE QUESTIONS (Page 115)
1. C [1]
2. D [1]
3. a) A broad absorption [1] in the range 3200–3600 cm^{-1} [1] for an O—H bond [1]. **[Allow an absorption in the range 1050–1300 cm^{-1} for 1 mark but no more than 3 marks in total]**
   b) The quantity of IR absorbed is proportional to concentration [1].
   c) Water [1], since it contains O—H bonds [1].
   d) Alcohol shows C—H absorption / absorbs at 2850–2950 cm^{-1} [1] but water does not [1].
4. a) The C═O bond [1]
   b) The bonds vibrate more [1].
   c) It absorbs infrared radiation from the Earth [1] so that less energy escapes to space [1].
   d) 3.080% [1]
5. a) There is peak at 2500–3200 cm^{-1} indicating O—H bonds in a carboxylic acid [1]. There is a peak at 1700–1750 cm^{-1} indicating the presence of a C═O group [1].
   b) A peak at 1700–1750 cm^{-1} [1].
   c) The peaks for carboxylic acids, alcohols and aldehydes are all very close together so may overlap [1]; all could be present in the product spectrum [1].

The Periodic Table

periods →
groups ↓

	1	2												3	4	5	6	7	0
																			4.0 **He** 2 Helium
	6.9 **Li** 3 Lithium	9.0 **Be** 4 Beryllium												10.8 **B** 5 Boron	12.0 **C** 6 Carbon	14.0 **N** 7 Nitrogen	16.0 **O** 8 Oxygen	19.0 **F** 9 Fluorine	20.2 **Ne** 10 Neon
	23.0 **Na** 11 Sodium	24.3 **Mg** 12 Magnesium												27.0 **Al** 13 Aluminium	28.1 **Si** 14 Silicon	31.0 **P** 15 Phosphorus	32.1 **S** 16 Sulfur	35.5 **Cl** 17 Chlorine	39.9 **Ar** 18 Argon
	39.1 **K** 19 Potassium	40.1 **Ca** 20 Calcium	45.0 **Sc** 21 Scandium	47.9 **Ti** 22 Titanium	50.9 **V** 23 Vanadium	52.0 **Cr** 24 Chromium	54.9 **Mn** 25 Manganese	55.8 **Fe** 26 Iron	58.9 **Co** 27 Cobalt	58.7 **Ni** 28 Nickel	63.5 **Cu** 29 Copper	65.4 **Zn** 30 Zinc		69.7 **Ga** 31 Gallium	72.6 **Ge** 32 Germanium	74.9 **As** 33 Arsenic	79.0 **Se** 34 Selenium	79.9 **Br** 35 Bromine	83.8 **Kr** 36 Krypton
	85.5 **Rb** 37 Rubidium	87.6 **Sr** 38 Strontium	88.9 **Y** 39 Yttrium	91.2 **Zr** 40 Zirconium	92.9 **Nb** 41 Niobium	95.9 **Mo** 42 Molybdenum	[98] **Tc** 43 Technetium	101.1 **Ru** 44 Ruthenium	102.9 **Rh** 45 Rhodium	106.4 **Pd** 46 Palladium	107.9 **Ag** 47 Silver	112.4 **Cd** 48 Cadmium		114.8 **In** 49 Indium	118.7 **Sn** 50 Tin	121.8 **Sb** 51 Antimony	127.6 **Te** 52 Tellurium	126.9 **I** 53 Iodine	131.3 **Xe** 54 Xenon
	132.9 **Cs** 55 Caesium	137.3 **Ba** 56 Barium	138.9 **La*** 57 Lanthanum	178.5 **Hf** 72 Hafnium	180.9 **Ta** 73 Tantalum	183.8 **W** 74 Tungsten	186.2 **Re** 75 Rhenium	190.2 **Os** 76 Osmium	192.2 **Ir** 77 Iridium	195.1 **Pt** 78 Platinum	197.0 **Au** 79 Gold	200.6 **Hg** 80 Mercury		204.4 **Tl** 81 Thallium	207.2 **Pb** 82 Lead	209.0 **Bi** 83 Bismuth	[209] **Po** 84 Polonium	[210] **At** 85 Astatine	[222] **Rn** 86 Radon
	[223] **Fr** 87 Francium	[226] **Ra** 88 Radium	[227] **Ac*** 89 Actinium	[261] **Rf** 104 Rutherfordium	[262] **Db** 105 Dubnium	[266] **Sg** 106 Seaborgium	[264] **Bh** 107 Bohrium	[277] **Hs** 108 Hassium	[268] **Mt** 109 Meitnerium	[271] **Ds** 110 Darmstadtium	[272] **Rg** 111 Roentgenium								

Key:
relative atomic mass — 1.0
atomic symbol — **H**
atomic number — 1
name — Hydrogen

Elements with atomic numbers 112–116 have been reported but not fully authenticated

lanthanides

140.1 **Ce** 58 Cerium	140.9 **Pr** 59 Praseodymium	144.2 **Nd** 60 Neodymium	144.9 **Pm** 61 Promethium	150.4 **Sm** 62 Samarium	152.0 **Eu** 63 Europium	157.2 **Gd** 64 Gadolinium	158.9 **Tb** 65 Terbium	162.5 **Dy** 66 Dysprosium	164.9 **Ho** 67 Holmium	167.3 **Er** 68 Erbium	168.9 **Tm** 69 Thulium	173.0 **Yb** 70 Ytterbium	175.0 **Lu** 71 Lutetium

actinides

232.0 **Th** 90 Thorium	[231] **Pa** 91 Protactinium	238.1 **U** 92 Uranium	[237] **Np** 93 Neptunium	[242] **Pu** 94 Plutonium	[243] **Am** 95 Americium	[247] **Cm** 96 Curium	[245] **Bk** 97 Berkelium	[251] **Cf** 98 Californium	[254] **Es** 99 Einsteinium	[253] **Fm** 100 Fermium	[256] **Md** 101 Mendelevium	[254] **No** 102 Nobelium	[257] **Lr** 103 Lawrencium

Index

acid anhydrides 102
acids 70, 73–4, 78
activation energy 48–9
addition polymers 98
air pollution 89
alcohols 93, 97, 100–3, 112
aldehydes 101, 112
alkanes 84, 88–95, 97–8
alkenes 93, 96–9, 101
alpha (α)-decay 64–5
amines 93
ammonia 74
ammonium ion test 78
anions 8, 10, 20–3, 61–2, 73–4, 77–8
anodes 61–2
aromatic compounds 85, 102
atom economy 53
atomic numbers 4, 69
atomic radii 69
atomic structure 4–7, 66

balanced equations 9–10
barium compounds 76, 78
beta (β)-decay 64–5
biobased/biodegradable polymers 82
biodiesel/bioethanol 80
boiling points 72, 88
bond angles 33–4, 96, 100
bond enthalpy 40, 45, 92
bonding electrons 24–5
bromine 73, 97

Cahn–Ingold–Prelog rules 96
carbocations 97
carbon 32–5, 92
carbonates 8, 78
carbon dioxide 114
carbon monoxide 88
carbon skeletons 84–5
carboxylic acids 101, 102, 112–13
catalysts 48–9, 57–8, 89
cathodes 61–2
cations 10, 20–3, 28–31, 61–2, 69–70, 74, 78
chain isomers 86
charge density 21, 70
chemical equilibria 56–9
chemical formulae 8–11
chlorine 73, 76
chlorofluorocarbons 94
collision theory 48
combustion 81–2, 88–9, 102
compound ions 8–11, 20

concentrations 16–18, 48–50, 57–8, 112
conductivity 21, 26, 28, 32–3, 69
copper sulfate 62
covalent bonding 24–7, 32–5, 85–6, 89–93, 96–8, 100
cracking 81, 88
cyclic compounds 85

dative (coordinate) bonds 24
delocalisation 28, 32–3
dependent variables 50
diamond 32
dilutions 18
dipole–dipole forces 36
displacement reactions 72–3
displayed formulae 85
disproportionation 61, 73
distillation 88, 104–5
double bonds 96

electrolysis 61–2
electron capture 64–5
electronegativity 69, 92–3, 97, 100–1
electronic configuration 6, 12–15
electron pair repulsion theory 33–4
electrons 4–7, 12–15, 24–5, 36, 66, 77
electron sub-shells 12–15
electrophiles 97–8
electrophilic additions 97–8
electrostatic attraction 20
elimination reactions 93, 101
empirical formulae 52
endothermic reactions 40, 56–8
energetics 40–3
enthalpy 40–7
environment 80–3, 113–14
equilibria 56–9
esters 102, 113
ethanol combustion 102
ethers 102
exothermic reactions 40, 56–8
experimental error 41
experimental techniques 104–7

flame tests 77
flue gases 76
fluorine/fluoride 72, 76–7
formulae 8–11
fractional distillation 88
free radical substitution 89–90
frequency 66
fuel cells 81
fuels 80–1, 88–9

fullerenes 32–3
functional groups 84–6, 112–15
fusion 66

gamma (γ)-decay 64–5
gases 53–4
global warming 113–14
graphene 32
graphite 32
green chemistry 82
group 1 compounds 70
group 2 elements 68–71, 76–9
group 7 elements 72–9

Haber process 58
half equations 10, 61
half-lives 65
halides 72–5, 77–8
haloalkanes 89–90, 92–5, 97, 101
halogens 72–9
Hess's law 44–7
heterogeneous catalysts 49
heterolytic bond fission 92–4
homogeneous catalysts 49
homologous series 84
homolytic fission 89–90, 94
hydrogenations 97–8
hydrogen bonding 37–8, 100
hydrogen energy 80
hydrogen halides 73, 74
hydroxide ion test 78
hydroxides 70

ideal gas equation 54
immiscible liquids 105
incineration 81–2
independent variables 50
infrared spectroscopy 112–15
initiation 89
inorganic tests 77–8
insoluble bases 22
instantaneous dipole–induced dipole forces 36
intermolecular forces 36–9
iodine 73
ionic bonding 20–3
ionic compounds 8–11, 20–3, 72–3
ionic equations 10, 72–3
ionic lattices 20–1
ionic precipitation 73
ionic radii 21
ionisation 13–14, 69, 108
ions 8–11, 13–14, 20–3, 34, 77–8, 108–11

isoelectronic species 20
isomers 86, 92, 96, 101
isotopes 4–5, 65

ketones 101, 112
kinetics 48–51

Le Chatelier's principle 56–8
limiting reagents 42
lone pairs 24, 33–4
Lyman series 66

malleability 28
mass numbers 4
mass spectrometry 108–11
Maxwell–Boltzmann distributions 48
medicine 65, 76
melting points 25–6, 28, 32, 68, 69, 72, 106
metal halides 70
metal ions 21, 77
metallic bonding 28–31
metals 68–71, 76–9
molar mass 16
molecular formulae 52, 85, 109
molecular ions 34
molecular mass 16
monoatomic ions 20
multiple covalent bonds 24–5, 34, 96

neutralisation 76
neutrons 4–7
nitrates 8, 70
nitriles 93
nitrogen oxide 89
non-metal ions 21
nuclear equations 65
nuclear reactions 64–7
nucleophiles 92–4
nucleophilic substitution 92–3, 101

organic chemistry 84–115
oxidation 60–3, 72, 101–2
oxidation numbers 60–2
oxides 70
ozone layer 94

parts per million 114
pascals 54
p-block elements 68
percentage uncertainty 46
percentage yield 52–3
periodicity 68–75

permanent dipole–induced dipole forces 36–7
pH 74, 76
phenols 102, 112
phosphates 8
photochemical smog 89
photons 66
pi (π) bonds 25, 96
plasticisers 98
polarity 25, 37–8, 92–3, 97, 100–1
pollution 89
polylactide 82
polymers 81–2, 98
positional isomers 86
positrons 64–5
precipitation 22, 73, 77–8
pressure 54, 57–8
products 44–7, 49–50, 56–8
propagation 90
protons 4–7
purity 17

quantities at equilibrium 57–8
quantum model of electrons 6, 12–15

radiation 64–7
radicals 89–90
radii
    atoms 69
    ions 21
radioactive decay 64–5
radiocarbon dating 65
rates of reaction 48–51
reactants 44–7, 56–8
reacting masses 16
reactive electrodes 62
reactivity 69–70, 72–3
recrystallisation 105
recycling 81
redox 60–3
reduction 60–3
refluxing 104
reforming 88
relative atomic masses 5
relative formula mass 16
rounding up/down 18

salts 22, 70
sigma (σ) bonds 25, 32–3, 96
silver halide test 77–8
skeletal formulae 85
smog 89
sodium chloride 62, 73

soil pH 76
solubility 22, 26, 28, 38, 70
solvents 26, 100
spectra 66, 108–15
spectrometry/spectroscopy 108–15
standard enthalpy changes 40, 44–7
stereoisomers 86, 96
sterilisation 76
structural formulae 85, 109–10
structural isomers 86
sub-atomic particles 4–7, 12–15
substitution reactions 89–90, 92–3
sulfates 8, 70, 78
sulfur dioxide 76, 89

temperature 40–1, 48, 54–8, 113–14
termination 89–90
thermal cracking 88
thermal decomposition 70, 74
thin-layer chromatography 106
titration 17, 22, 29–30

van der Waals forces 36
variable oxidation states 61
variables 50
volume of gases 53
volumetric solutions 29

water 16–17, 38, 76–7, 114
wavelength 66, 77

yield 52–3

Acknowledgements

The author and publisher are grateful to the copyright holders for permission to use quoted materials and images.

Cover & P1: © Yellowj. Shutterstock.com

All other images are © Shutterstock.com and © HarperCollins*Publishers* Ltd

Every effort has been made to trace copyright holders and obtain their permission for the use of copyright material. The author and publisher will gladly receive information enabling them to rectify any error or omission in subsequent editions. All facts are correct at time of going to press.

Published by Letts Educational
An imprint of HarperCollins*Publishers*
1 London Bridge Street
London SE1 9GF

ISBN: 9780008179090

First published 2016

10 9 8 7 6 5 4 3 2 1

© HarperCollins*Publishers* Limited 2016

All rights reserved. No part of this publication may be reproduced, stored in a retrieval system, or transmitted, in any form or by any means, electronic, mechanical, photocopying, recording or otherwise, without the prior permission of Letts Educational.

British Library Cataloguing in Publication Data.
A CIP record of this book is available from the British Library.

Series Concept and Development: Emily Linnett and Katherine Wilkinson
Commissioning and Series Editor: Chantal Addy
Author: Alison Dennis
Project Manager and Editorial: Tanya Solomons
Index: Indexing Specialists (UK) Ltd.
Cover Design: Paul Oates
Inside Concept Design: Ian Wrigley and Paul Oates
Text Design, Layout and Artwork: Q2A Media
Production: Lyndsey Rogers
Printed in Italy by Grafica Veneta SpA

MIX
Paper from responsible sources
FSC™ C007454